Guide to Highway Law for Architects, Engineers, Surveyors and Contractors

T0188207

Guide to Highway Law for Architects, Engineers, Surveyors and Contractors

Robert A. O'Hara
Senior Lecturer in Construction Law
School of Financial Services and Law
Sheffield City Polytechnic

Taylor & Francis
Taylor & Francis Group

LONDON AND NEW YORK

Published by Taylor & Francis
2 Park Square, Milton Park, Abingdon, Oxon OX14 4RN
52 Vanderbilt Avenue, New York, NY 10017

First issued in paperback 2020

Taylor & Francis is an imprint of Taylor & Francis

First edition 1991

© 1991 Robert A. O'Hara

Typeset in 9/10pt Palatino by Just Pages, Sussex

A catalogue record for this book is available from the British Library

Library of Congress Cataloguing-in-Publication data available

Publisher's Note
The publisher has gone to great lengths to ensure the quality of this reprint but points out that some imperfections in the original may be apparent

ISBN 13: 978-0-367-58007-0 (pbk)
ISBN 13: 978-0-419-17330-4 (hbk)

Contents

Glossary   viii
Section 1 Introduction   1
Meaning of Highway   2
Types of Highway at Common Law   3
Creation of Highways   3
   (i)   By Dedication   4
   (ii) Other Aspects   5
   (iii) By Statute   6
Highways Legislation   6
The Highways Act 1980   7
Administration of the Act   7
Other Acts of Parliament   8
Highway Authorities   9
Classification of Highways   9
Maintenance and Improvement of Highways   10
Modern Maintenance   11
   (i)   Maintainable at Public Expense   11
   (ii) Private Maintenance   12

Section 2 Rights of Adjacent Owners   14
Miscellaneous Rights   14
The Vesting of Sewers   15
Obligations of Adjacent Owners and or Occupiers   16
Remedies for Public Nuisance   17
Rights of the Public   17
Limitations of the Rights of the Public   18
Access   19
Carriage Crossings   21
Stopping up a Means of Access   22
Stopping up, Diversion and Extinguishing Highways   22
   (i)   The 1980 Act Procedure   23
   (ii) The 1971 Act Procedure   25
Protection of Highways   27
   (i)   Criminal Offence at Common Law   27
   (ii) Common Law Nuisance   29
   (iii) Criminal Offence under Statute   31
   (iv) Construction Operations near the Highway   32

Section 3 Building Operations on or near the Highway 33
Activities Related to Construction Work 33
   (i)   Work which causes Actual Damage to the Highway 33
   (ii) Prohibition of Soiling the Highway 33
   (iii) Power to Remove Structures and Projections 34
   (iv) Safety Precautions to Prevent Damage to the Street 35
   (v) Building close by, under or over a Highway 35
   (vi) Facilitating Building Operations 36
Builders' Skips 37
   (i)   Defences 38
   (ii) Removal of Builders' Skips 39
   (iii) Approved Owners 40
Hoardings 40
Scaffolding 41
'Lines' 42
   (i)   Building Lines 43
   (ii) Improvement Lines 43

Section 4 Making up New Streets 45
'New Streets' 46
Town and Country Planning 49
Enforcement 49
Private Street Works Code 50
Advance Payments Code 51
   (i)   Exemptions to the Code 51
   (ii) Amount of Payment 53
   (iii) Power to Insist on Making up Road 53
Agreement under S38 53
Private Roads 54
Adoption of Streets 55
   (i)   At the Discretion of the Authority 55
   (ii) 'Democratic' Action 55
   (iii) Agreement under S38 56
Serving of Notice 56

Section 5 Civil Liability 57
The Building Owner 57
The Contractor and the Sub-Contractor 59
The Construction Professional 61
The Local Highway Authority 62

Section 6 Short Case Studies 65

Appendices   123

1   Problems and Questions   123
2   Table of Cases   127
3   Table of Statutes   131

Bibliography   135

Index   137

Glossary

The following abbreviation of law reports are used in the text.

A.C.	Appeal Cases
ALL E.R.	ALL England Reports
B.L.R.	Building Law Reports
Burr	Burrows Reports, Kings Bench
Ch.	Chancery Division
Crim. L.R.	Criminal Law Reports
E & B	Ellis and Blackburn Reports, Queens Bench
E.G.	Estate Gazette
I.R.	Irish Reports
J.P. (L.R.)	Justice of the Peace Law Reports
L.G.R.	Local Government Reports
L.T.	Law Times
P. & C.R.	Planning and Compensation
Q.B.D.	Queens Bench Division
R.T.R.	Road Traffic Reports
Term	Term Report
T.L.R.	Times Law Report
Vent	Ventris Report
W.L.R.	Weekly Law Report
W.R.	Weekly Report

Section 1

Introduction

At a time when so much construction work is being carried out on or near to highways, a contractor ignores at his peril the law of highways and the influence it has, or should have, on his working methods and practices; should he do so not only might he incur claims for damages from members of the general public whose rights he might infringe in his ignorance, but he may also face criminal charges laid by the local highway authority or police which may result in fines being imposed, and or the serving of injunctions (this may be either civil or criminal).

Some knowledge, understanding or awareness of the law relating to highways is essential to anyone involved in the construction process, including the architect, engineer or surveyor advising a client as to what is possible, and the contractor actually carrying out the contract works; this can be illustrated by say the influence an improvement line might have on a building design, or where access eg a carriage crossing to a property might be positioned or whether to attempt to obtain a section 38 agreement or use the advance payments code when building new roads; the contractor also would be influenced in many ways by such things as restrictions on rubbish skips parked on the highways, lorries leaving his site spreading mud on the road and heavy vehicles damaging the highway eg breaking kerb stones and paving slabs. These examples are simply elementary illustrations to help emphasise the importance of some knowledge of the law of highways.   These instances relate to practical examples involving the actual building work itself but they are not the only kind of liability that arises from working on or near the highway.

The law of highways imposes on land owners adjacent to highways a strict liability regarding injuries to members of the general public resulting from work carried out on their land even if the work is carried out by independent contractors, Tarry v Ashton (1876), (1) this also helps to highlight the role of the professional when advising a client regarding a project and its possible legal implications. From such a short description of some of the difficulties which may arise it is apparent that for each and everyone involved in the construction process, from design through to the work on site a knowledge of the law relating to highways is essential if the project is going to be fulfilled satisfactorily; all professionals should be aware of the law within which they must operate, the opportunities it offers and the constraints it imposes; it is to this end that this book is directed.

Meaning of Highway

Highway law has a long history stretching back to the middle ages, and has developed mainly through the common law, and during the last century, increasingly and more importantly through legislation, to give us the law applicable in the present day; an indication of its detail, volume and complexity can be ascertained by considering that the Highways Act 1980 (the principal Act) consists of 347 sections and 25 schedules, which must make it one of the longest pieces of legislation on the statute book.

The law of highways is unique in that it grants public rights over private property so that members of the general public have a certain basic right to use a highway, the landowner retaining proprietary rights and in circumstances when that right is abused, being able to obtain a legal remedy against members of the general public.

The most commonly quoted definition is: 'a highway is a way over which there exists a public right of passage, that is to say a right for all Her Majesty's subjects in all seasons of the year freely and at their will to pass and repass without let or hindrance...' Ex Parte Lewis (1888) 21 (2); We must remember that the right is limited to 'pass and repass' and has been so for hundreds of years as can be illustrated from Goodtitle and Chester v Alker and Another (1757) (3) that'... in a highway a king hath but passage for himself and his people...', and also that the landowner retains full ownership and control, notwithstanding the existence of a highway subject only to any applicable statutory provisions, as is shown by a second statement from the same case (Goodtitle) that the landowner'... has the right to all above and underground except only the right of passage for the king and his people...'.

A highway is not of necessity simply a roadway or carriageway but also may be a footpath, bridleway or driftway. A highway may be limited to a particular type of use Hue v Whitely (1929) (4);

with limitations being placed on the use of the highway, this occurring most commonly with footpaths, bridleways and driftways. The essential characteristic of a highway is that all members of the general public have 'the right' to 'pass and repass' so that a highway cannot be dedicated to a limited section of the public such as say, the inhabitants of a particular parish, street, or locality. If a way comes under private control it is not and cannot be a highway because use would not be 'as of right' but only with permission of the owner, 'on sufferance or by licence'.

Types of Highway at Common Law

There are three kinds of highway according to the limits of passage over them:

1 A cartway or carriageway is a highway over which the public has a right of way: (i) on foot, (ii) riding or accompanied by a beast of burden, (iii) with vehicles and cattle (with the possibility of the latter being excluded).

2 A bridleway is a highway over which the rights of passage are limited by excluding vehicles and sometimes the right of driftway (driving cattle), they are limited to foot traffic and the riding or leading of horses depending on limitations imposed and advertised by signposts; also it now includes the riding of bicycles, which became lawful in these circumstances following the enactment of the Countryside Act 1968 S30.

3 A footpath is a highway where the right of passage is limited to passage on foot.

Creation of Highways

The creation of a highway comes about either through the common law doctrine of dedication and acceptance or by the application of a statutory provision. In theory a public right of way may be claimed on prescription, (the vesting of a right by reason of lapse of time - originated in Roman law) ie continuous uninterrupted use for a period of 20 years Fairey v Southampton CC (1956) 2, (5); but it would be defeated on proof that the right came about after 1188 (time immemorial). As user by the public may be evidence of dedication it is not necessary nowadays to rely on common law prescription as 20 years' established use will bring about the dedication under statute S31 1980 Act, ie there is an 'implied' dedication.

Highways, then, come into existence in one of two ways:

(i) by dedication
(ii) by statute.

(i) By Dedication

A road or way becomes a highway when the owner of the soil either expressly or by inference from his conduct indicates that the public should have a right of passage ('right to pass and repass') across his land. The acceptance is acknowledged by the public using the way (referred to as 'user by the public' or 'general public user'); and from the moment that a dedicated way has been accepted by the public that right 'to pass and repass' comes into existence and a highway is created.

Express dedication comes about when by formal deed to the local highway authority the owner of the soil grants the right and the local authority accepts; the most striking example of this being when a developer completes the construction of estate roads, under the supervision of the authority, eg a S38, 1980 Act agreement, and hands them over to the local authority upon which they become highways and in these circumstances fully the responsibility of the authority for maintenance and repair.

An important point to note is that under common law the dedication of itself does not create a highway (the above example is conducted under statutory rule) but that there must be acceptance by the general public (public user) 'as of right'.

Dedication presupposes animus dedicandi, ie an intention to dedicate; as above it may be expressed in writing or words but (outside statutory arrangements) is often, if not usually, a matter of inference brought about by regular public use, ie implied dedication. This will be so where the owner has either said things or conducted himself in such a way as to imply that he has granted the right of way to the public, A-G v Esher Linoleum Co Ltd (1901), (6); and that the dedication may be inferred from a time prior to the earliest proved user, Williams and Ellis v Cobb (1935), (7), where evidence of use prior to 1856 was admitted to establish dedication (similarity with Fairey's case.)

To prevent the inference being drawn the landowner must in some way show he has no intention of dedicating the way; it must be done in some clear, positive manner or by legal process; it is insufficient for him to claim he has no intention to dedicate, ie that the intention not to dedicate was 'locked away in his head' or that he told a person, a total stranger to the locality, or turned back a stranger on an isolated occasion. Consequently notices clearly indicating a lack of intention to dedicate require to be posted eg 'private way', 'Not Public Right of Way' etc. These challenge the assumption that continuous use has existed for 20 years and defeat S31 1980 Act. The same effect is obtained by a landowner by giving notice to the local authority and indicating on a map deposited with it, those ways over his land which are highways and those which are not.

A landowner can show his control, ie that the right to use the way is under 'licence or sufference' by interfering with the way and preventing its use; for this purpose it is usually satisfactory if the way is closed one day a year Lewis v Thomas (1950) (8). (Traditionally the railway companies provided many examples of this as many 'ways' were on railway land over which they intended keeping control.)

Other Aspects
1 There is no public right to wander at will ('jus spatiandi'). This could only be granted by statute. The use of an esplanade for strolling up and down or for amusement is not inconsistent with it being a highway: Sandgate UDC v Kent CC (1898, (9); Ramus v Southend Local Board (1892), (10), where the promenade became a highway. It must not though become or be a nuisance: A-G v Blackpool Corporation (1907), (11): where motor racing was held to be illegal. (Note present day developments in Birmingham where a Local Act has granted authority for motor racing on the ring road and equally with cycle racing in city centres.)
2 Cul-de-sacs There is no rule of law which prevents a cul-de-sac from being a highway but the defining of a way as a highway usually infers 'somewhere to go to' and is unlikely to be inferred as such simply by public usage. However if there is something of interest to see at the end of the cul-de-sac that may be sufficient to justify creation of a public highway. Roberts v Webster (1967), (12); Moser v Ambleside UDC (1925), (13).
3 Foreshore There is no general public right of way over the foreshore, but a highway may exist along or across a particular part of it.
4 Artificial Structures A Highway may exist over embankments, sea walls and piers. Greenwich Board of Works v Maudsley (1870), (14).

5 Churchways. This is a way over which parishioners, only, have a right of way to go to and from their parish church. It is distinguished from a highway by the fact that the right is applicable to a restricted class, ie the parishioners and not members of the general public, it is impossible to create in modern times: Farquhar v Newbury UDC (1909), (15). In A-G v Mallock (1931), (16); the public have no right to enter a church at all times: dedication is not inferred.

(ii) By Statute
The Highways Act 1980, as other previous Highways Acts, has provisions whereby either the Minister or the Local Highway Authority can obtain land, either by agreement or compulsorily, to build roads which automatically, without public user, become highways maintainable at public expense. This would as a matter of course include any drainage of the highway that may be necessary. Should the statute authorise merely the setting out and making of a road it fails to become a highway automatically until it is set out or substantially completed but should there be public user of incomplete roads, dedication and acceptance may be inferred: Cubitt v Lady Caroline Maxse (1873), Hals, (17).

Previously it has been alluded to that legislation has increasingly become more important in this area of law and it is consequently worth while to consider this development in some detail.

Highways Legislation

Legislation has for centuries had an influence on the law of highways, one of the earliest examples being from the Statute of Winchester 1285 whereby bushes and undergrowth had to be cleared for up to two hundred feet on either side of the highway in an attempt to make it less convenient in providing shelter for robbers (shades of modern under-passes and the problems they create!) Up to the end of the 18th century there were many different statutes applicable to highways, usually introduced to meet a particular need and only local in application, ie local bye-laws; the same principle still applies in that Local Acts contain powers applicable only to that particular local government area, eg Sheffield, Manchester, Leeds, Birmingham etc. Towards the end of the 18th century many of those Acts were consolidated and amended, for example in 1766 and 1773 and remained in force until reforming legislation of the 19th century was enacted, this becoming known as the Highways Act 1835 to 1885 and remaining the main legislation in this field until the Highways Act 1959, this Act being the basis of today's legislation. The 1959 Act was subsequently amended in 1961 and 1971, the whole being repealed and replaced by the Highways Act 1980 which came into force on the 1st January 1981 and is now the principal Act.

The Highways Act 1980

The 1980 Act is an extremely long and detailed piece of legislation consisting of 345 sections and 25 schedules dealing with all aspects of highway work much of it impinging directly on the work of the contractor which he should take into consideration when planning layouts and production programmes, for instance providing access to a site, S184, or the temporary closing of a highway to facilitate either demolition or unloading of materials (obstruction S136) or the co-ordination in a production programme of a local authority team to break open the highway and make a sewer connection, while other parts are an influence only on the periphery of the actual building process as in which code to use when 'making up' streets Part XI, and is more the concern of the professional adviser to the client than the contractor.

The Act provides rules and procedures governing the construction of highways from legal creation and construction specification, to the control of the behaviour of people using the highway be they building owners, contractors or members of the general public (S 168 (1)). This latter point should not be confused with the obligations imposed on the use of vehicular traffic by the Road Traffic Acts.

Administration of the Act

The Act makes provision for the creation of highway authorities and agreements between these authorities to initiate, supervise and regulate the many provisions and powers set down in the Act. The following are only some of the provisions made: the creation of trunk roads, special roads; maintenance and improvement of highways; stopping up and or diversion of highways, stopping up of access to highways; laying out and the construction of new streets, the making up of private streets; to regulate the use of builders skips (S139) on the highway and a whole range of building activity from mixing mortar to stacking materials, erecting scaffolding (S169) and hoardings (S172); other powers regulate the positioning of buildings by designating improvement lines and building lines as well as powers to order the reverse hanging of doors, and gates which open outwards over the highway (S153).

There are also sections dealing with what are called 'savings' (S333 to S339) whereby the section(s) state clearly that the Act does not automatically authorise a particular action but exceptions can be made by the appropriate authority (S336); permission to extract materials from the foreshore obtained under the Coast Protection Act (1949) (S18, S34) or by completion of the correct procedure (as with Telegraph Act 1878 S7 or S338 '80 Act which shows the 1980 Act does not interfere with the Post Office's (or British Telecom's) rights under the Telegraph Acts 1878).

Other Acts of Parliament

Other Acts of Parliament have an influence on work in or on the highway as in the following examples:

The National Parks and Access to the Countryside Act 1949 and the Countryside Act 1968 both have provisions relating to highways in the form of footpaths and bridleways which have little relevance to builders other than in the event of obtaining a diversion related to development.

The Town and Country Planning Act 1971 contains important provisions for the stopping up and diversion of highways (S209 to 221) so as to enable development to be carried out; also it enables the removal of vehicular traffic as in shopping or pedestrian precincts or 'malls' (S212); the Act also makes provision for compensation in the event of property being 'blighted' by the plans of a proposed highway (S 192-208); a carriage crossing is also considered to be development for which a planning permission is required (S22(1), 23) Town and Country Planning Act 1971; this work comes within the definition of Engineering Operations (S290 of the '71 Act).

The Public Utilities Act 1950 contains a code to be followed by the public utilities (water, gas, electric, telephones) when carrying out work in the street by 'breaking open' the highways or sometimes in land adjacent to the street where it is maintained by the highway authority as land reserved for a prospective highway.

The Building Act 1984 grants to local authorities powers to deal with dangerous buildings or structures and to make certain requirements of owners to improve the paving in respect of certain courts, yards or passages and entrances to them (S84-85) – these may be highways under the Highways Act 1980.

Highway Authorities

The administration of the Act is carried out by what are called Highway Authorities set up under the Act consisting of the Minister of Transport and sometimes the Secretary of State of the Environment (in Wales the Secretary of State for Wales) to carry out government functions under the Act whereas functions at local level are carried out by County Councils and District Councils; following the abolition of the Metropolitan County Councils and the GLC which were highway authorities, highway functions revert to the District Councils in those affected areas which now become highway authorities. To distinguish them from the Minister these authorities are known as 'local highway authorities' or LHAs.

Highway authorities have the responsibility to create, construct and maintain highways in their own areas and to supervise and regulate by conditions, within their discretion, many of the different activities including building work which are carried out on or near to highways.

Parish Councils (in Wales, Community Councils) are not highway authorities, but do have some functions relating to highways; for example, its consent is required before a highway is stopped up, diverted or declared no longer maintainable at public expense. It is also within the powers of a parish council to pay for the maintenance of any highway maintainable at public expense within its area provided of course that it keeps within its spending resources.

Classification of Highways

Highways are classified as follows:

1 Special Roads These are roads reserved for particular classes of traffic, for example motorways with limitations on which type of vehicle and those drivers who may use them. The Minister is the highway authority.

2 Trunk Roads These are major roads which primarily but not exclusively carry through traffic in towns and cities with the Minister as the highway authority but with the possibility of this being transferred to the local highway authority.

3 Classified Roads These originally came about under the Ministry of Transport Act 1919 and it was important in that they

were the basis of the grant system from central to local government for the repair and maintenance of roads. Roads were classified into 'A', 'B' and 'C' roads with principal roads (similar to class 'A') being created in the Local Government Act 1966. There is little modern relevance to this distinction as money is now provided on a different basis by central government, ie through loans and grants to local authorities or as part of general rates and 'rate support' grant.

4 Unclassified Roads Any road which does not fall into any of the above categories.

5 Public Footpaths These are 'as of right' passage on foot only.

6 Bridleways These are as above either riding a horse or leading one; riding a cycle is also now lawful on a bridleway.

Maintenance and Improvement of Highways

Historically the responsibility for the repair and upkeep of highways was that of the people resident in the parish'... unless it could be shown that it was the responsibility of an individual or corporate body by reason of tenure, inclosure or prescription...' and also 'if it be a public way, of common right the parish is to repair it unless a particular person be obliged by prescription or custom...' Per Hale J. Austins Case (1672), (18).

To fulfil this collective obligation of the parish local people had to be prepared to give time to maintain the local highways. Making the population as a whole responsible was nothing new in mediaeval times, for example under the Statute of Winchester 1285 the community, 'the hundred', were made collectively responsible for all crime in their area and were made collectively liable for punishment for failing to discover the perpetrators of known crimes. This same statute introduced (more in hope than anticipation?) the idea of clearing brush-wood from either side of the highway between towns (referred to on p.6). In practice little was done to ensure correct maintenance but the Statute of Highways 1563 imposed a repair liability and allowed the appointment of a 'surveyor of highways' chosen by the parish who was required to impose six days unpaid compulsory labour a year on all inhabitants who were occupiers of land. Under the 1691 Highways Act the surveyor was appointed by the magistrates and subject to direction by them; he was also empowered to levy a sixpenny rate to defray the expense of any work in excess of the six days compulsory free labour. 'The inhabitants at large' were responsible and could be prosecuted at quarter sessions for failing to meet even the low standard of repair demanded, but the liability was a criminal one as opposed to a civil one as illustrated in Russell v Men of Devon (1788) 2 Term Rep 667, (19), where it was held that there was no civil liability for personal injury caused by a badly maintained highway, but towards the end of the 19th century injury caused by negligent repair (mis-feasance), ie positive acts of negligence, rather than failing to repair (non-feasance) ie not carrying out any repairs, was being recognised as in Thompson v Brighton Corp (1894), (20), where the duty was recognised but the plaintiff lost, but in Skilton v Epsom and Ewell UDC (1937), (21), the plaintiff was successful: but the defence was still successful in Wilson v Kingston upon Thames Corp [1949], (59):

Modern Maintenance
The Highways Act 1835 brought highways legislation into the modern era; parish highway authorities were formed which fulfilled the functions previously carried out by the magistrates and the compulsory unpaid labour burden on the parish was removed. The 1835 Act also introduced the concept of 'adoption' regarding the public responsibility for maintenance and repair. Prior to the Act all highways were automatically publicly maintained by the parish or 'the inhabitants at large'. This Act changed this so that highways that existed prior to the Act would remain a public responsibility but that roads built subsequently would only become maintainable at public expense when 'adopted', ie taken over by the local authority which would only occur when the roads were 'made up' to the surveyor's satisfaction; this Act also created for the first time certain highways for which no one was responsible for maintenance and generally known as 'private streets' (p.50) as distinct from 'private roads'. It is worth noting that the word 'road' is used not 'highway' in the Act so consequently footpaths and bridleways remained as before, the responsibility of 'the inhabitants at large'. This remained so until the National Parks and Access to the Countryside Act 1949 when certain provisions of the 1835 Act were applied to newly created ones.

Highway maintenance falls into two main categories: public authority, and private

(i) Maintenance at Public Expense

The liability to maintain which the legislation has imposed (S41) applies only to highways which were maintainable before the 1835 Act and all public carriageways adopted by the local highway authorities since that time. It also includes footpaths and bridleways which existed prior to 1949 and any created later if adopted. New roads constructed following development work 'vested' in this way, ie adopted by agreement as in S36 to 40 of the '80 Act. If a local authority builds a highway on its own land it automatically becomes maintainable at public expense.

(ii) Private Maintenance

Highways are maintained privately by reason of:

(i) an individual or corporate body being responsible for maintenance based on a statute,
(ii) prescription ie long-established practice,
(iii) land tenure,
(iv) enclosure,
(v) owners of land on either side of a private street.

(i) An individual or corporate body may be liable to repair under a statute but today it is a rare obligation but examples from case law can be used to illustrate the point. The Macclesfield Corp v Great Central Railway (1911), (22), 'GCR' did not repair a highway on a bridge over a canal; it was the company's obligation under statute (the railways and canals were built under the authority of Act of Parliament which included maintaining highways which crossed canals, and railway tracks). Macclesfield Corporation repaired it and sued to recover the costs of repair; the court decided that it was 'GCR's' responsibility but the corporation had volunteered to repair it and could not recover the cost.

(ii) By prescription where there is evidence that repairs have been carried out over a long period of time, ie from time immemorial, this can be defeated by showing that the highway originated after that time unless the courts allow a claim of 'lost grant' as it may do if the landowner concerned has been receiving a consideration, eg taking a toll.

Rundle v Hearle (1898), (24):

(iii) by reason of tenure brings with it the liability for repair if it is established that it arises resulting from tenure of the land; it will run with the land and it has to be shown that the tenants or predecessors have carried out repair works from 'time immemorial'; this is similar to prescription and can be defeated in a similar manner, ie to show that it was created within legal memory.

(iv) By reason of enclosure, whereby a landowner can enclose, ie fence off, usually on both sides of the highway, so as to control wandering or deviation of animals. It may be that several ways have been established and this deviation can be removed by fencing, leaving a way of reasonable width which brings with it a liability for repair of the highway. Without evidence of deviation the liability to repair is avoided as it is if permission is obtained from the LHA under S51. Any liability to repair as a result of fencing is avoided if the fence is later removed.

(v) The landowner of a private street is not liable in law for its maintenance (it was one of those classes of highway for which no-one was made responsible following the 1835 Act) but should the circumstances demand it he may be served with a notice from the LHA to carry out urgent repairs (S 230 1980 Act).

Section 2

Rights of Adjacent Owners

Where a highway is dedicated to the public or presumed to be so at common law, the owner of the land retains the property in the soil and may even convey it or leave it to others. In the case of a highway maintainable at public expense only the 'skin' ie the formation of the road including drains, (colloquially referred to as 'two spits' deep) is vested in the highway authority. Tunbridge Wells Corporation v Baird (1896), (25), the public's right is of course only one of passage, ie the 'right to pass and repass'.

At common law there is a presumption, subject to any evidence to the contrary, that the owner of land adjoining a highway owns the soil under the highway up to its centre line; also at common law the owner has the right to tunnel under, bridge over and to create access to his land at any point where his land abuts a highway. Such a right must be exercised with reasonableness regarding the rights of the general public. Tottenham UDC v Rowley (1912), (26). However, modern legislation has limited these rights with statutory permissions being required before work could start on any of the above examples. If as an example we look at access to land adjoining a highway and the way it is now controlled by statute, we gain some insight into the way legislation works. (p. 19 and 21.)

Miscellaneous Rights

The ownership of trees and herbage (grazing rights) lie in the owner of the soil and any problems ensuing, eg roots growing into drains or foundations of adjacent buildings or raised footpaths are the responsibility of the landowners public or private. In the case of highways maintainable at public expense it is assumed the local authority would accept liability even when they do not own the soil beneath the highway! The landowner has no right to break open the highway without a statutory permission but should he do so unlawfully and provided the highway is made good satisfactorily, then any pipes or cables which may have been laid and had been the reason for excavating the highway in the first place, may remain in position. The owner of the soil has many rights at common law as previously mentioned; he also has authority over the highway and may restrict the erection of wires above the highway; these rights are now subject to modern statutory control.

The consent of the owner of the soil is required before people, other than statutory bodies acting under statutory rules, may lay pipes, dig tunnels etc; a present day example may well be the insertion of cables into the highway in preparation for cable television if this was not done under statutory authority. Three cases illustrate the concept of statutory permission.

Marriot v East Grinstead Gas and Water Co. (1909), (27). The water company was successfully sued for trespass for laying pipes without landowner's permission, under a privately maintained footpath (a highway) without the landowner's permission.

Wood v Ealing Tenants (1907), (28); 'W' connected his drain to a pipe which drained more than one property and was therefore a sewer. It was vested in the LA making it a public sewer.
To make the connection 'W' laid his pipe in land belonging to 'ET' without permission. 'W' trespassed when laying pipe.

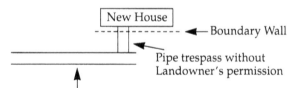

Public sewer in private land

Porter v Ipswich Corporation (1922), (29); 'IC' erected two poles 6' 0' deep in highway to carry electric power cables; 'P' as owner of soil, sued 'IC' for trespass. Held that the LA did not request his permission. Action failed: there was a proviso regarding permission but it applied only to highways not dedicated to public use, and maintained at public expense.

The Vesting of Sewers. Tunbridge Wells Corporation v Baird (1896), (25). Under the PHA 1875 certain streets were vested in the LA. In a Local Act the authority were empowered to build public toilets. When the LA excavated for the toilets it was challenged by 'B' who claimed that the soil was not vested in the authority even if the Act gave it 'control of the streets'. The LA did not own the soil beneath the highway and lost the case as a result of not requesting permission from the owner of the soil.

Obligations of Adjacent Owners and or Occupiers

It is considered wrong that a person should create a danger on his land and if injury is then caused to a user of the highway, claim as a defence that the source of the danger was on private property. This can be illustrated in Fenna v Clare (1895), (30); where a child was injured by falling onto spikes on the top of a low wall and also in Harold v Watney (1898), CA (31); where a child was injured climbing a rotten fence at the side of the highway. The person involved must be using the highway; it will not apply if the person is on the other side of the fence to the highway, Bromley v Mercer (1922), (32).

Damage caused by natural projections over the highway is not actionable unless the danger was apparent or should have been so to a reasonably prudent landowner or occupier as in Caminer v Northern and London Investment Trust (1951), (33); where the landowner was not liable when a tree with decaying roots (unknown to the landowner) uprooted and fell on to a passing car, following the decision on similar facts in Noble v Harrison (1926), (43).

In the event of man-made projections the liability is strict as in Tarry v Ashton, (1); where an outside lamp fell on a visitor. If the danger has not been caused by the landowner this is a good defence but should it not be dealt with as soon as its existence becomes known or should have become known (presumed knowledge of nuisance) ie to allow the danger to continue, then the landowner becomes liable as in Leasne v Lord Egerton (1943), (34); where a broken window-pane fell on to a passer-by three days after the damage was caused by an air raid.

It is then a public nuisance if owners or occupiers of property abutting the highway allow their property to fall into disrepair so that it is a danger to users of the highway or that it may deter people from using it. This would also apply if bushes or trees were allowed to grow over the highway.

There is a requirement in common law that fences should be

erected and or maintained only if some artificially created danger exists eg building and works, excavations etc; someone who is using the highway may accidentally deviate from it and be injured and consequently deter others from using the highway. Crane v South Suburban Gas Company (1916 , (35); (regarding distance from highway case law suggests 50 yards is not excessive, but it must depend on circumstances).

Remedies for Public Nuisance

Every public nuisance is at common law an offence triable either summarily or on indictment. If at the time of the trial the nuisance still exists the court may order the defendant to remove it or stop it (an 'injunction') ie to abate it; no lapse of time and no consent other than under statute affords a defence to public nuisance. Damages as a remedy are available if the plaintiff can show special damage as in Halsey v Esso (1961), (36).

The law also allows a person injured by a nuisance an element of self help ie 'abatement of a nuisance' but only within strictly defined limits. The owner of the soil may remove an obstruction where a person is obstructed from exercising his public rights by some obstruction on the highway. However abatement is a remedy the law does not favour and is usually not advisable. Lagan Case (1927), (37); it is a right to remove the cause of the nuisance and no more; it is not a licence to inflict unnecessary damage upon the individual responsible or to exact retribution. Neither does it allow for possible actions in the future, for example, removing scaffolding from a neighbouring building site for fear that the house to be built will infringe a right of light.

Legal action to obtain an injunction to restrain the commission of a nuisance on the highway or an instruction to remove one may be instigated by the Attorney General, also local authorities (without his assistance) may take legal action to protect the interests of the local population.

Rights of the Public

These are restricted to the 'right of passage' only, this has been compared with an easement in common law Dovaston v Payne (1795), (38), but is denied in modern and law other kinds of conduct by members of the general public may infringe the property rights of the owner of the soil giving him a remedy in trespass to land.

The 'right of passage' is then for legitimate travel only; for it not to be so or for it to be conduct which could be defined as unreasonable or not a proper user of the highway, will create the basis of an action for trespass by the adjacent owner. What is 'reasonable' or a 'proper use' of the highway can only be defined by case law and the following give some indication of what is not acceptable.

In Hickman v Maisey (1900), (39), 'M' who from the highway was observing 'H' train racehorses to obtain better information for placing bets was held to be a trespasser as this conduct was considered to be not a legitimate use of the highway.

The concept of reasonableness is important in the common law and as user of the highway the general public must conduct itself so; reasonable user is illustrated in the Harrison v Duke of Rutland (1893), (40), where 'H' was from his position on the highway interfering with a grouse shoot, he was held to the ground by two gamekeepers following instructions from the Duke whom he sued. 'H' lost, Lord Esher saying 'Highways are no doubt dedicated prima facie for the purpose of passage; but things are done upon them by everybody which are recognised as being rightly done, and as constituting a reasonable and usual mode of using a highway as such'. Accordingly, people could rest or sketch for a reasonable time. In Rogers v Ministry of Transport (1952) 1, (41), it was considered a reasonable use of the highway for lorry drivers to park on the grass verge whilst in a cafe for a meal; it was a temporary stop, the grass verge was part of the highway and there was no obstruction.

Highway authorities have the power and sometimes the duty (S 130) to 'assert and protect the rights of the public to use and enjoy the highway' R v Surrey CC, Ex p Send Parish Council (1979), (42).

Limitations on the Rights of the Public
The general public have the right of passage and this can only be limited in the way discussed above. It can though be controlled and influenced in another way which is that of preventing deviation or in setting clear boundaries. Where there are no clear specific boundaries the width of the highway will be a matter of fact unless stated in legislation such as the Inclosure Acts. The existence of metalled roads does not of itself delineate the line of the highway as strips of land on either side may form part of the dedicated highway. In a similar way land up to fences or up to ditches (privately owned boundary to private land) is presumed to be part of the highway.

Where there is no defined path across a field the right seems confined to a path of reasonable width running straight from terminus to terminus subject only to a right of deviation if the direct path is impassable.

Access

Every landowner adjoining a highway is entitled to have access to his property from the highway; this is a basic common law right as is amply illustrated in Tottenham UDC v Rowley, (26); Marshall v Blackpool Corporation (1935), (44) and Cobb v Saxby (1914), (45). These cases show that the landowner is entitled to access for both foot and vehicular traffic, subject of course to any statutory constraints which may exist. The present legislative rules on access are to be found in the Town and Country Planning Act 1971 and in the Highways Act 1980. Also Perry v Stanborough (Developments) Ltd and Wimborne DC and Dorset CC (1977) Journal of Environmental Planning Law (46), is a cautionary tale for when purchasing land as also is R v Baker (1980), (23): things should never deteriorate so!

The right of access is a private property right the infringement of which would create a right of action on behalf of the owner and the possibility of his obtaining a legal remedy such as damages or an injunction. In the Blackpool case the infringement of a private right was where the corporation was held to have acted in excess of its statutory powers, ie ultra vires, in withholding permission for the building of a carriage crossing; whereas in Cobb v Saxby (1914), (45) it was where one shopkeeper prevented his neighbouring shopkeeper from having access to the wall of his own shop which he used for advertising purposes.

Other examples would be activities that might be considered as an unreasonable use of the highway and as such an interference with the property rights of the owner as in Barber v Penley (1893), (47) where crowds outside a theatre prevented 'B' gaining access to his premises or as in Leonidis v Thames Water Authority (1979), (48) where road works prevented access to business premises. Other examples of unreasonable use of the highway would be someone repairing cars on a business basis on the roadside or people parking vehicles deliberately across an entrance to prevent the owner gaining access.

The actions complained of in these cases are known as private nuisance and are an infringement of the private property rights of the owner. Should the actions complained of be an interference with public rights then these would constitute a public nuisance which is a crime and as such is enforced, within their discretion, by the authorities which would be either the local highway authority or the police. Should the actions complained of be a public nuisance, the adjacent owner is required to show special damage, ie damage localised to himself, over and above that suffered by the public at large before he would be able to obtain a civil remedy. Therefore if an adjacent owner, say a shopkeeper, suffers a loss as a result of building operations, for example regular parking of heavy vehicles unloading precast concrete units or the parking of ready-mix concrete trucks, the action would be based on obstruction of the highway, a public nuisance, ie breach of a public right, but with the shopkeeper claiming special damages for loss of profit for the duration of the obstruction; these kinds of claims were made in Lyons, Sons & Co. v Gulliver (1914),(49) and Harper v Haden (1933), (50).

Should the action complained of be as a result of authorisation under a statute no action will lie and no remedy will be available, unless the statute authorises it. However an action may be successful if the conduct complained of is done oppressively or in such a way as to cause unnecessary damage as in Millward v Reddith Local Board of Health (1873), (51).

Actions which may constitute a public nuisance may be carried out by adjacent owners also, the most common examples being where a shopkeeper encroaches on the footpath to display goods or where someone may repair motor vehicles at the roadside. Contractors may very easily commit a public nuisance with scaffolding or hoardings erected on a footpath, the careless parking of vehicles for unloading purposes, or the stacking of materials on the footpath. All of these examples, and there are many more, could constitute a public nuisance with the consequences outlined. A further serious point requiring emphasis is that under S137 of the Highways Act 1980, the police may arrest without warrant anyone who '... in any way wilfully obstructs the free passage along the highway...' The fine is a maximum of £150.

The fundamental approach to be adopted when working in circumstances where a nuisance may be committed is to work in a reasonable manner considering the rights of both the neighbours and the general public. If work has to be done which creates these problems, and in construction work it is inevitable that there be disruptions in other people's activities, then awareness of the possibilities is required at the planning stage of the project so that full cognisance can be taken of the law and its consequences, also that any permissions that may be required are obtained, so making the building activities lawful.

Carriage Crossings

One of the most common ways of obtaining access with vehicular traffic to a property is by what is called a carriage crossing whereby a special ramped approach is constructed from the carriageway across any grass verge and footpath to the boundary of the property, so providing a continuous smooth link between the carriageway and the property and so avoiding damage to the footpath (and/or the grass verge) and vehicles alike. Both the positioning and construction are subject to statutory control under the Town and Country Planning Act 1971 and the Highways Act 1980 respectively.

The provision of a 'means of access' to the highway is part of the definition of 'Engineering operations' under S290 (1) of the 1971 Act and as such is classed as development, (S22 '71 Act) therefore requiring a planning permission before work on site can start. It applies to 'any means of access whether private or public, for vehicles or for foot passengers and includes a street...' The main emphasis of a planning permission for a means of access is one of road safety; the positioning of a vehicle crossing will depend on the relative safety or otherwise of the requested position with permission being granted for the crossing away from dangerous corners or blind bends. Conditions may also be imposed regarding the materials used (S29), for example, tarmacadam or granite setts etc. Hawkins v Minister of Housing and Local Government (1962), (52); also with larger buildings conditions may be imposed regarding forecourts, slip roads etc to avoid any unreasonable interference with the flow of traffic.

The local highway authority will examine the construction, ie technical details and eventually either carry out the work itself and charge the landowner an appropriate fee or if permission is given for the owner to construct his own access, supervise the work in progress; this supervision is important because these works become part of the highway maintainable at public expense and so therefore the authority will make certain the work is to an acceptable standard before accepting that liability. Similar provisions can be applied (S127 1980 Act) to existing crossings if they are situated in potentially dangerous situations, so that the local highway authority may request the re-siting of a crossing to a less dangerous position, always assuming that this is possible, if not the landowner cannot be denied access to the highway (except only in the most extreme circumstances (see below p.22) and the crossing would remain. In addition to the powers outlined above the LHA may, where there has been regular use across a footpath by vehicular traffic, insist on the construction of a crossing at the owners' expense (S184 1980 Act). Continuing to cross the footpath in this manner without a crossing is an offence (S184 (17), carrying a fine of up to £50 on a second offence.

Stopping up a Means of Access

The means of access to a highway from adjoining land can be 'stopped up', ie legally removed, and physically closed, under two sets of circumstances: (i) where the access creates a danger to passing traffic or is likely to interfere unreasonably with it (S 124 1980 Act) or (ii) where a planning permission has been granted for road improvements under S211 of the 1971 Act, whereby one highway is prevented from intersecting with the new improved highway; they are 'stopped up'. In both cases the 'stopping up' is brought about by either the local highway authority or the Minister of Transport issuing an order to that effect. The order thus issued is subject to two conditions: (i) that other reasonable access is available or will be provided by the authorities (either Minister or local) and (ii) that the confirming authority (usually the Minister or a LHA order), must be satisfied that no access to the premises from the highway is reasonably required (S124 1980 Act.) These two examples come about as a result of a specific order to 'stop up' but access may be cut off or prevented as a result of road improvements (S76 1971 Act) as for example when a corner is 'banked up' creating conditions where the road surface is several metres or feet higher than that of the original surface (this was never allowed under the common law rule) having the effect that all vehicular access from the properties on that corner is made impossible. In each of these circumstances where the party concerned can show that a loss has been suffered, as for example a reduction in market value of a property, then compensation may be claimed; any dispute over entitlement or amount is heard by the Lands Tribunal. All that has been stated above presupposes disputes and conflicts and creating procedures to resolve them but it need not always be so. Section 127 1971 Act does envisage voluntary agreements between the local authorities and the adjoining owner.

Stopping up, Diversion and Extinguishing Highways

The common law rule of 'once a highway always a highway' illustrates a principle, long established, that in contrast with the creation of a highway, which often came about on the basis of implied dedication or prescription, over a period of time, there is no way in which the public can reverse the process and release the rights acquired and no authority can bind the public in purporting to release those rights. There are ways in which a highway may be extinguished, as when for example public access at both ends is cut off by lawful stoppng up or by natural causes as when land slips occur or there is erosion caused by the sea. Historically the only method of legally extinguishing a highway was after a hearing by a jury following the issue of a writ of 'ad quod damnum' 'to what damage' (not to damage public interest): the jury had to decide whether the proposed closure would be detrimental to the public. This procedure is the one established in common law but which has now been superseded by statutory procedures to meet modern requirements.

In recent times with the constant growth and development of our built environment and the ever changing face of our towns and cities it became important that these developments be not unnecessarily held up or delayed because of legal difficulties encountered in wishing to develop sites or construct buildings which interfered with existing highways. It is to this end that procedures to enable stopping up, extinguishing or the diversion of highways under a variety of statutes were introduced. The Highways Act 1980 and the Town and Country Planning Act 1971 are the Acts which contain the procedures most commonly used. Other Acts make similar provision, such as the Housing Act 1985, but contain far more specialised requirements as in section 294 of the 1985 Act where a local authority can make an order extinguishing highways on land acquired for clearance and development; also the Civil Aviation Act 1949 (S28) enables the Minister of Transport to order the stopping up or diversion of any highway if 'it is necessary to do so in order to secure the safe and efficient use of land for civil aviation purposes...' Similar powers exist to extinguish public rights of way over land which has been obtained by compulsory purchase in the Acquisition of Land (Authorisation Procedure) Act 1946; oddly this applies only to ways not used by vehicular traffic.

(i) The 1980 Act Procedure
Under the 1980 Act (S116) a magistrate's court may by order authorise the 'stopping up' or diversion of a highway if the court (at least two magistrates) is convinced that the highway 'is unnecessary' or that the diversion requested would be 'nearer and more commodious to the public...' There are qualifications in that the application has to be made by an 'appropriate authority' ie the local highway authority and the power does not apply to trunk or special roads. Should someone other than the authority wish to obtain an order of this nature, say a landowner or estate developer,

they may request the local authority to make the application as the 'appropriate authority' (S117); the authority, in acceding to the request may make it conditional, eg that all expenses incurred by the application be defrayed by the developer. If trunk or special roads were involved in a scheme it would be necessary to obtain an order, from the Minister under S10 and S325, providing that the road in question ceases to be a trunk or special road and bringing it within the purview of S116.

The decision to 'stop up' a highway is made by the magistrates court which must before arriving at a decision ensure that certain parties have been informed and have been given the opportunity to make representations to the court as set down in Schedule 12 Part 1 of the 1980 Act. This states that certain organisations and interested parties have to be informed, eg district councils, parish councils, adjacent landowners and that certain acts of publicity have to be carried out namely that the notice of the application has been published in the London Gazette and also in at least one local newspaper and that notices have been posted at either end of the way affected; this gives notice and enables anyone concerned to prepare and present objections to the court. There is time for this as there has to be 28 days' notice of the hearing. Anyone not satisfied with the decision can appeal it to the Crown Court (S317 1980 Act).

Whilst S116 refers to 'highways' nevertheless S118 to 121 deal specifically with 'stopping up', extinguishing and the diversion of footpaths and bridleways. Whereas under section 116 the decision to extinguish or divert is made by a court, under S118 the decision, specifically related to footpaths and bridleways, is made by the local authority being confirmed by the Secretary of State for the Environment, ie the first decision is a judicial one whereas the latter is administrative in nature. Under S118 a council may make an order called a 'public path extinguishing order' to bring about the end of the public right of way. The basis on which it is made is if it appears to the 'council... that it is expedient that the path or way should be 'stopped up' on the ground that it is not needed for public use...' In a similar manner if it is to secure' the efficient use of land or of other land held with it or providing a shorter or more commodious path or way it is expedient...' that the line of footpath or bridleway be diverted. These decisions are arrived at by the council and the Minister only after the procedure set down in Schedule 6 of the 1980 Act has been complied with. In both cases all of the people likely to have an interest must be notified of the proposals being made and a procedure set out in Schedule 6, 1980 Act (which is similar to Schedule 12 Part 1 p.23) must be complied with. Up to this point the procedure is more or less common to both types of application, but regarding the decision making process there is a difference in the preparation for making the decision in that in the case of a 'public path extinguishing order' if objections are not received then the authority can make the order and itself confirm it. Only if there are objections is the Secretary of State required to confirm the order. In the case of a 'public path diversion order' if objections are made and not withdrawn, the Minister must hold an enquiry; there is a choice: it may be a local inquiry or it may be one by someone appointed for the purpose of giving objectors an opportunity to make representations. Should a local authority be an objector, then a local enquiry must be held. After the enquiry the Minister may confirm the order with or without modifications. In the case of diversions, should an owner or occupier of land be able to establish as fact that a loss has been sustained, then it is possible that a claim for compensation under S28 might be successful, any disputes being settled by the Lands Tribunal.

(ii) The 1971 Act Procedure

Commonly known as the 'planning method' this procedure is always associated with a planning application or a planning permission granted under Part III of the 1971 Act; the powers and the procedures to be followed are contained in sections 209 to 221.

The procedures of the 1980 Act with regard to stopping up etc are rather time-consuming, so to speed up the process the 1971 Act procedure allows decision by Ministerial Order rather than having to make application to the magistrates court, such powers being accommodated by Section 123 (1) 1980 Act, where it states 'the foregoing provisions... do not prejudice any power confirmed by any other enactment [ie 1971 Act]... to stop up or divert a highway...' In this way different procedures, with similar objectives, from various Acts of Parliament exist side by side. Orders may be made by either the Secretary of State or the Local Planning Authority depending upon which section an order is being applied for.

The procedures are carried out by two distinct bodies: the Secretary of State for the Environment and the Local Planning Authority. Often they work independently of each other; on other occasions they work together, with the Secretary of State either confirming, or not, a decision by the authority. They each have their own role to play in making orders and then, where necessary, in confirming them. Should an order be required following the granting of a planning permission it is for the Secretary of State to grant it having considered that 'it is necessary to do so in order to enable development to be carried out in accordance with the planning permission granted...' The Secretary of State makes those orders (the whole process is governed by S215 1971 Act) which become necessary following the granting of a planning permission and also draft orders (S216) which may be made and which run concurrently with the progress of a planning application, but can only become final after the granting of the planning permission in itself again reducing what might be not inconsiderable delay. It is also within the Secretary of State's purview to grant an order 'stopping up' and diverting highways which intrude on to a new highway or a greatly improved one which has received planning permission from the local planning authority, and it is considered expedient to do so in the interests of safety and efficient traffic flow (S211). Similarly where a local authority may wish to convert existing carriageways into 'shopping malls' or 'shopping precincts', requiring limiting the traffic to pedestrian use only, or if not, restricting the use of vehicles to certain times of day, this can only be done (S212) by order of the Secretary of State following consultations with the local planning authority. It would be an order extinguishing the right of people to use vehicular traffic on the highway. Following this, S213 allows the authority to improve the amenities available by creating flower beds etc.

The role of the local planning authority in making orders is limited to those affecting footpaths and bridleways (S210) which may require stopping up or diverting to enable a development to proceed, which order would require confirmation by the Secretary of State only if the making of the order was opposed; if not the local authority confirms the order. The procedure which is followed and the provisions of S210 are similar to what has to be done in the same circumstances under the 1980 Act, ie the publicity that has to be given to the proposals, the notification of various authorities and persons with a legal interest, eg landowners. The detail is contained in Schedule 20 1971 Act. When a local authority has acquired land to develop, it can issue an order extinguishing all public footpaths and bridleways (S214) provided that alternative ways have, or are to be, provided. For other highways to be extinguished it is necessary for the Secretary of State to make the order (S214 (1) (a) based on the same 'alternative way' approach. It is a quick and expeditious method of closing a highway.

What might happen and often does is that as a result of the actions mentioned above, individuals may suffer a loss. Parliament being aware of this has created the possibility of an aggrieved person being able to obtain compensation under section 212 (5) (6). It must be claimed within six months of the order and comply with the Town and Country Planning Regulations 1976 (Reg 14) SI 1976 No 1419.

Protection of Highways

Previously it has been shown that there exists only the right 'to pass and repass' regarding the use of the highway by the public at large and that this right has to be exercised in such a manner that it will be deemed by the courts to be a 'reasonable user' of the highway as in Rogers v Ministry of Transport (1952), (41). Unreasonable user of the highway is conduct which can be construed as being an interference with the highway in such a manner as to prevent members of the general public from exercising their right 'to pass and repass', ie an obstruction. Legal protection of the highway and its lawful users is provided by both the common law and statute, the nature of which may be either criminal or civil (in certain instances it is possible for both conditions to apply); therefore to use the highway without lawful authority or to interfere with it may constitute: (i) a criminal offence at common law, (ii) a common law nuisance (private or public), (iii) a criminal offence under statute (primarily the Highways Act 1980).

(i) A Criminal Offence at Common Law

This refers to any action which results in the highway being interfered with or obstructed; it is a crime (referred to as a public nuisance) unless carried out under some lawful authority or is conduct which can be defined as a 'reasonable user' of the highway, examples of which might be: obtaining a statutory permission to park a builder's skip on the highway and someone parking a motor car outside his house. Obstruction to the highway occurs regularly as for example when excavations take place to lay new sewers, telephone lines etc or when repairs are carried out to the carriageway surface or where builders erect scaffolding and hoardings during building work which project on to the footpath and therefore interfere with, ie obstruct the highway. These actions would be public nuisances punishable by fine unless carried out by lawful authority. These examples would require statutory authority as laid down in the Public Utilities Street Works Act 1950 and the Highways Act 1980.

For an obstruction to exist it does not require that there be some insurmountable obstacle in the way of the public exercising its right on the highway, nor that someone has been obstructed but it is sufficient that members of the public are denied free access to the

whole of the highway. This is clearly stated in Wolverton UDC v Willis (1962), (53); where a greengrocer obstructed the highway with a display projecting 11 in from the shop front and 12 ft 8 in long on a footpath over 6 ft 6 in wide. It was held that:
1 Every member of the public is entitled to unrestricted access to the whole of the footway.

2 Subject to the de minimus principle (discussed in Seekings v Clarke (1961) (54); the projection in this case was considered substantial) any encroachment on the footway and which restricts him and which is not authorised by law is an unlawful obstruction.

3 Every member of the public so restricted is necessarily obstructed in that he is denied access to the whole of the footway.' However what is 'reasonable user' may on the face of it seem like a complete obstruction as with a procession, but nevertheless be held to be reasonable as was emphasised in Lowdens v Keaveney (1903), (55).

It is not possible to avoid obstructing the highway at all times. It is done every day and to a certain extent has to be tolerated, for example: vehicles break down, vehicles park to load and unload, builders sometimes erect scaffolding and hoardings. If such things are necessary and reasonable even though they cause inconvenience or injury they will technically be obstructions but not public nuisances; only if actions of a similar nature are carried out carelessly or unreasonably does 'give and take' cease and they become criminal offences, ie public nuisances.

Examples of what constitutes a public nuisance are many and varied. The following are a selection to illustrate the point: causing queues to form; carrying out excavations; causing things to project which are a danger to pedestrians or vehicles, ie branches of trees or lamps; leaving dangerous or slippery substances on the highway; dangerous cellar flaps; smoke endangering passing motorists; hosepipe trailing across roadway; building materials stacked on the footpath.

To summarize: for an obstruction to be actionable it is not necessary for it to be dangerous as any obstruction is a clear infringement of the public's right of passage but nevertheless it must also be an unreasonable use of the highway.

(ii) Common Law Nuisance

The previous section dealt with obstruction being a common law crime, the remedy for which is fines or in extreme cases imprisonment; this section deals with the other element in public nuisance which allows an injured party to claim a remedy if it can be established that the individual suffered 'special or peculiar' damage over and above that of the public at large. Public nuisance is not as is private nuisance dependent on property rights, but is founded on the 'unreasonable interference with the comfort and convenience of life of a class of Her Majesty's subjects in the exercise of their common rights'. (1)

There are two forms of actionable nuisance:

    (i)   obstruction of the highway, and

    (ii) something which is a danger to highway users. It may not obstruct the highway but does create the potential for danger to users of the highway. It may even be on private land adjacent to the highway as in Castle v St Augustine's Links (1922), (56); which can be compared with Bolton v Stone (1951), (57); and Millar v Jackson (1977), (58) see also Fenna v Clare (1895), (30). The details of Tarry v Ashton (1), would suggest that there is strict liability on the owner or occupier (person instigating work complained of) of land abutting the highway but this seems related only to a complete failure of the work in hand rather than to the occurrence of some incident, though whilst causing damage, is subsidiary to the main work being carried out. This is shown in Padbury v Holliday and Greenwood (1912), (60); where a hammer knocked from a windowsill caused injury but was held to be the negligence of the contractor and not the liability of the owner; it was 'collateral negligence', 'before a superior employer could be held liable for the negligent act of a servant or sub-contractor it must be shown that the work which the sub-contractor was employed to do was work the nature of which, and not merely the performance of which, cast on the superior employer the duty of taking precautions.'

If the employer is to be liable the danger must be inherent in the work; it is not enough that the contractor (or sub-) chose a negligent manner to do the work where a normal method would create no foreseeable risk to the plaintiff. This followed a similar decision in Reedie v London and North Western Railway Co (1849) (61); where masonry was carelessly dropped on a passer-by during the building of a bridge. It seems that to

(1) (Stephen, Digest of Criminal Law art.235).

follow the rule in Tarry v Ashton the window would have to fall out or the bridge to collapse! But should a landowner allow a nuisance to continue he would be strictly liable as in Leasne v Lord Egerton (1943), (34). (also ref. 16.)

The claim for special damages usually relates to either loss of profit or to personal injuries. With loss of profit it is difficult to establish what would have been earned and to quantify the loss but successful claims have been made as in: Barber v Penley (1893) (47); people outside the Globe theatre following the earlier performances of 'Charley's Aunt', the queue was an obstruction and the theatre management were liable. A similar decision with an injuction being served on the theatre management, was made in Lyons Sons & Co. v Gulliver (1914), (49).

Where damages are sought the courts have to strike a balance between competing interests in coming to the decision as to whether or not the act complained of was a reasonable use of the highway. They will have regard to the status of the defendant, the manner of the creation of the nuisance, the physical nature of the obstruction and the length of time it has existed and finally the degree of fault on the part of the defendant. (This latter point is open to challenge and controversy as the cases are not clear whether fault is required.)

To be successful a claim of damages must relate to damage suffered other than that suffered by the public at large and the loss suffered must relate directly to the nuisance itself, ie the nuisance caused the accident. In Dymond v Pearce (1972), (65); a vehicle had been parked on the roadside for an unreasonable time in the dark without lights, but nevertheless the motor cyclist who injured himself when he collided with it was held to be riding his motor bike negligently and that this was the cause of the accident, not that the vehicle was causing an obstruction.

Should a user of the highway behave in a negligent manner then with that 'fault' will lie liability, so that for example contractors working on the highway owe a 'duty of care' to other users and bear the consequences should it be breached. The following are some cases which illustrate the point (1): Haley v London Electricity Board (1965), (66); Ellis v Sheffield Gas Consumers (1853), (67); Holliday v National Telephone Co (1899), (68).

1. Refer to Section 5, p.57

(iii) A Criminal Offence under Statute

Under the Highways Act 1980 many forms of obstruction and interference with the highway have been codified and are now also statutory offences. This makes the offences more specific and clearly defined so that (hopefully) it is possible for members of the public to know more clearly their rights and obligations in any given set of circumstances.

Many of a contractor's day to day activities come under the influence of this statute. For example: to carry out excavations in the highway, S131; to erect a scaffolding on or over the highway, S169; to park a builder's skip on the highway, S 139 (and many more) are all statutory offences if carried out without lawful authority (usually obtained from the local highway authority). Statutory offences may also be offences at common law under the general heading of obstruction or interference as in the examples above (and on p. 29) or they may be offences which are specifically statutory (ie actions which are offences simply because the statute says so and Parliament considers it necessary), as for example refusing to reverse the hanging of a door or bar which opens directly over the highway, S 153, or to use a hoarding which is 'not securely fixed to the satisfaction of the council' S173.

The penalty usually imposed is a fine which may vary, the statutory section breached laying down the maximum fine to be imposed and also giving authority where offenders are persistent to impose a daily fine for each day the offence continues; also in these circumstances the (l). See authority can apply for a mandatory injunction which would instruct the contractor to stop the conduct about which the complaint was made.

A serious situation may arise with what the Act refers to as 'wilful obstruction' S137; in the event the offender may be arrested without warrant by a police officer; also mens rea is not necessary for the offence to be committed; Arrowsmith v Jenkins (1963), (69) where 'A' was imprisoned for creating an obstruction by addressing a public meeting on the highway. Other Acts of Parliament which may be invoked might be the Town Police Clauses Act 1847, S28, and the Public Health Act 1875 S 171 (l).

The local highway authority has a duty to protect the right of the public to enjoy the highway and to prevent as far as possible the stopping up and obstruction of those highways for which it is responsible. The authority will be the body most likely to prosecute and obtain injunctions against offenders. (l)

(iv) Construction Operations near the Highway

If during the course of carrying out building operations in or near a street and as a result of which there is an accident with risk of serious personal injury to people using the street or an accident occurs which would have had a similar result had not the local authority (or LHA) used its emergency powers for dealing with dangerous buildings (Building Act 1984 S78), then the owner of the land or building is, without prejudice to any other liability, guilty of an offence, the maximum fine for which is £500.

A defence to this charge would be that he took 'all reasonable precautions to secure that the building operation was so carried out as to avoid causing danger to persons in the street, or that the accident 'was due to the act or default of another person (eg the contractor? or stranger) and that he took all reasonable precautions and exercised all due diligence to avoid the commission of such an offence by himself or anyone under his control'. The defence of 'another person' can only be used if evidence as to the identity of that person or evidence which will assist in his identification is served to the prosecutor at least seven clear days before any hearing to decide the case. Without this evidence this defence is unavailable without leave of the court.

Under JCT 1980 cl 20 the contractor agrees to indemnify the employer 'against any expense, liability, loss, claim proceedings whatsoever arising under any statute or at common law in respect of personal injury to or death of any person whomsoever arising out of or in course of or caused by the carrying out of the works...'

This condition seems on the face of it to provide a safeguard for the employer, ie the building owner, should someone be injured in the street resulting from the works. This clause is effective from the point of view of any civil claims which an injured party may bring but even though it specifically mentions 'any proceedings... under any statute or the common law', this breach is a breach of statute but is also a crime for which a fine may be imposed. The question arises whether in these circumstances, as opposed to civil actions, the clause would provide protection for the employer in terms of reimbursement for legal costs and the amount of the fine imposed. Should events take this turn and the contractor refuse to indemnify in this way, could the employer enforce the clause? As the clause appears to seek to oust the jurisdiction of the courts it would seem to be not enforceable.

Section 3

Building Operations on or near the Highway

Activities Related to Construction Work

The 1980 Act influences and in many instances provides legal controls of many specific operations associated with construction work on or near the highway. Whilst it is impossible to discuss in detail each and every section of the Act which might apply in these circumstances, it is possible to look in some detail at many of the more common building operations which are affected and also those for which the outcome in event of infringement would be most serious from the point of view of both the construction programme with its attendant delays and the legal effects in terms of fines and or damages; Part IX of the Act contains the majority of the sections, the breach of which could have this result.

(i) Work which causes Actual Damage to the Highway

These sections (132, 133, 134) reflect concern for the physical state of the highway prohibiting without permission, as they do, any excavations and damage to footways and also providing a remedy when the sites of excavation on land adjacent to the way collapse; Lodge Hole Colliery v Wednesbury Comp (1908), (70).

There is also the prohibition of, and the power to remove, any unauthorised marks on the highways (S132); this is intended to control graffitti and unofficial signs but might apply to the contractor if levels used for setting out on site were established on the road surface adjacent to site. Statutory undertakers are prevented except in emergencies from excavating the highway if during the previous twelve months the road has been closed to traffic or the width has been reduced to less than two thirds for the purpose of carrying out roadworks and repairs.

(ii) Prohibition of Soiling the Highway

In an attempt to keep the highway clean it is prohibited to deposit on the highway anything that is a nuisance with S148 specifically prohibiting dung, rubbish,and material used for dressing land (probably aimed at farmers muck spreading). The spreading of mud on the highway would come under this section if it could be shown that in the circumstances it is a nuisance; to avoid this occurring it is good management practice to ensure that there is provision, usually a water tap and hosepipe close to the exit from a site so as to enable drivers to wash down the wheels of their vehicles. The deposit of mud on the highway is often specifically dealt with by local byelaw under the Local Government Act 1972; enforcement, though, can be difficult as a result of the confirming authority, ie, the Department of Environment, refusing to permit the inclusion in the byelaw of a provision requiring the owner of the vehicle (often apparent or easily obtained) from disclosing the name of the driver.

Section 150 deals with the duty imposed to clear snow and soil washed onto or which falls onto the highway, whereas S151 gives powers to instigate works, by notice to occupier, to prevent soil being washed onto the highway.

Rainwater should not be allowed to simply 'run off' buildings onto the highway, there should be some provision (S163) of gutters, downpipes and channels to remove surface water to the drains.

Trailing or stringing wires, ropes, pipes or other apparatus across a highway is forbidden under S162: Trevitt v Lee (1955), (71); Clarke v J Sugrue & Sons (1959), (72); Farrell v Mowlem (1954), (77).

(iii) Power to Remove Structures and Projections

These sections cause certain actions to be carried out: to get things done. Under section 143 it is an offence to erect structures in the street without lawful authority and these may be removed by the local highway authority - included in the definition under S143 is '...machines, pumps, posts or other objects...' this could involve the contractor using the highway to park a compressor or pump or to, insert posts to form a barrier or fence around, say, an excavation.

The taking down of projections which might present a danger to the public can be enforced under S152 with strict liability being imposed on the occupier following Tarry v Ashton should personal injury be the result of such danger. Should '...doors, gates or bars...' open onto the street the authority can insist under S153 that they are reversed (windows are not mentioned) whilst under S164 barbed wire, if a nuisance, must be removed from fencing adjoining a highway; Stewart v Wright (1893) (73); barbed wire fixed at a level of less than 8 feet (2.400m) above the highway is an offence under S28 Town Police Clauses Act 1847.

(iv) Safety Precautions to Prevent Damage to the Street

The metalled surface of the street itself is protected by the prohibition of certain activities either on the street or on any land nearby. The use of firearms, setting off fireworks, the lighting of fires are all forbidden under S161; if they occur within 50 feet of the centre of the highway and results in injury to or interrupt a user, or endanger a person using the highway then an offence is committed with a maximum penalty of £50.

The setting of fires by builders involved in demolition work can lead to this offence being committed as also can stubble burning by farmers and bonfires on 5th November.

Allowing dirt, filth or lime to run or flow onto the street warrants a fine of £10 which is also the penalty for playing football or any other games in the street (beware lunchtime football matches on site), several cases illustrate the issues raised: King (Contractors) v Page (1970), (74); Gatland v Metropolitan Police (1968), (75); Hunston v Last (1965), (76); Crane v South Suburban Gas Co (1961), (35).

As previously mentioned dangerous activities on land adjacent to a highway are influenced by the 1980 Act, S165 deals with the consequences of the land itself being dangerous for example unfenced excavations, or dangerous dilapidated fencing or excavations that through weathering encroach up to the highway. Cases which discuss some of these aspects are: Carshalton UDC v Burrage (1911), (78); Myers v Harrow Corp (1962), (79); Nicholson v Southern Railway Co (1935), (80).

Continuing the same theme, S166 allows the authority to issue a notice requiring the owner or occupier to carry out such work as is specified in the notice so as to remove the danger, obstruction or inconvenience to the public of any steps or projections on a forecourt.

(v) Building close by, under or over a Highway

The highway cannot be interfered with without authority; consequently should someone erect a building or fence or plant a hedge in a highway which consists of or comprises a carriageway, an offence is committed (S138). A distinction is made between a highway with a carriageway and one without; therefore footpaths and bridleways are excluded from the scope of this section. The planting of shrubs is prohibited 15 feet (4.500 m)from the centre of the carriageway (S141) but a licence may be granted to an owner or occupier to allow shrubs, trees etc to be planted; this could be useful where the architect wishes to landscape the approaches to a building or housing estate.

To prevent landslips and rubble and rubbish from falling onto a highway retaining walls are often built and any that come within 12 feet (3.600 m) of the highway and are to be over 4 ft 6 in (1.350 m) high require permission from the local highway authority which will be granted when the authority is satisfied regarding the constructional detail of the wall following the submission of detailed drawings (this is not necessary if the transport authorities are the land owners).

Restrictions are imposed on the erection of any structure which spans the highway or projects over it unless lawful authority is granted (S176, 177, 178), this would include bridges, beams, wires, rails etc. New Towns Commission v Hemel Hempstead (1962), (81). Similarly when buildings are constructed over the highway, Pilling v Abergele UDC (1950), (82).

Under S286 the authority can require the external angles of a building to be 'rounded off'.

To build a cellar under the highway without authority is an offence and should one be so constructed the authority can take steps to remove it (S179). The provision of natural light and ventilation to cellars and the maintenance of them plus that of the roof where the cellar extends under the street is controlled by S180 and a case which discusses the rights and obligations of a landowner who retains responsibility for cellars in this position is provided in: Principality Building Society v Cardiff Corporation (1968), (83).

(vi) Facilitating Building Operations

The Act recognises that it is virtually impossible to construct a building on land which abuts a street without in some way obstructing or interfering with that highway. Consequently the Act allows the contractor to work on the carriageway and to conduct his operations with such things as scaffolding and hoardings that would otherwise without authority be classed as obstructions. It is permissible without gaining permission (S170) to mix mortar on the highway if it is mixed on a 'plate' or in some receptacle; it is an offence to 'mix mortar or deposit or mix on a highway cement or other substance... likely to stick to the highway... or if it enters the drains or sewer... is likely to solidify'. The fine is a maximum of

£200, local authorities and statutory undertakers are excluded from the effects of this section. It is possible to obtain a licence (S180) to allow mixing if it cannot be reasonably done elsewhere (cable layers for cable TV may come into this category).

To be able to deposit materials on the highway when either preparing for a contract or when fulfilling one and then to tip rubbish there during construction and on completion can be a distinct advantage to a contractor, S171 allows this following a granting of permission which may also impose conditions: Drury v Camden LBC (1972), (84).

A duty is placed on the employees of local authority (S175) to take all reasonable precautions to ensure that materials which are left on the highway overnight with their authority, are guarded or left in circumstances, lights, warning notices etc where the risk of accident is minimised, failure to do so may result in a personal fine of up to £25 and possibly liability in negligence following personal injury: Hardcastle v Bielby (1982), (85).

Builders' Skips

The use of skips or hoppers to remove waste from building sites large and small is one of the most common publicly observed activities of builders today and often these skips require parking on the highway as the site itself is too small or restricted; as is often the case with refurbishment and house extension contracts: it seems sometimes that there is a builder's skip parked around every corner of our suburban streets.

Under the 1980 Act S139, skips may be parked on the highway following the granting of permission by the local highway authority. Should a skip be deposited without permission the owner of the skip is guilty of an offence and liable to a fine not exceeding £100. The authority has discretionary powers to impose conditions on a permission, a list of such conditions is set down in S139 (1) as follows:

(a) the siting of the skip,
(b) its dimensions,
(c) the manner in which it is to be painted to make it visible to oncoming traffic,
(d) the care and disposal of its content,
(e) the manner in which it will be lighted or guarded,
(f) its removal at the end of the period of permission.

When a permission is granted there are certain mandatory conditions which must be fulfilled by the owner (S139(4)), failure to do so may incur a maximum fine of £100; they are as follows:

(a) the skip is lighted properly during the hours of darkness (half an hour after sunset and half an hour before sunrise (S329 (1) Saper v Hungate Builders; King v Hungate Builders (1972), (86): Lambeth BC v Saunders Transport (1974), (87),

(b) that the skip is clearly and indelibly marked with the owner's name, telephone number or address,

(c) that the skip is removed as soon as practicable after it has been filled,

(d) that each condition subject to which that permission was granted is complied with.

Where offences are committed under this section they often arise out of the conduct of others: for example where at night lamps are stolen or simply thrown into the skip, creating a hazardous obstruction on the highway, or the driver of the transport lorry may simply 'drop' (deposit) the skip in the place most convenient to himself ignoring the conditions contained in the permission.

The responsibility is essentially that of the owner but not exclusively so; also statutory defences are provided under S139.

(i) Defences

Essentially there are three offences under S139, related to the depositing of a 'skip':

(i) without permission under S139(3)

(ii) and not fulfilling the statutory conditions of S139(4)

(iii) which is 'obstructing or interrupting any user of, a highway' S139(9).

There does not appear to be a specific offence if LHA conditions are infringed as opposed to S139(4) ones, whether to do so invalidates the permission so as to make it an offence to deposit or to continue to deposit a 'skip' is not made clear.

As a consequence of the permission to deposit a 'skip' on the highway being granted to the owner (defined in S139) any charge which is brought regarding any infringement of S139 is brought against him. Two main defences are granted by S139(9):

(i) that the deposit was as a result of a statutory permission S139(3) and that all the statutory conditions laid down in S139(4) have been complied with,

(ii) that the offence was committed by another person and that he (the owner) had taken all reasonable precautions and exercised all due diligence to avoid the commission of such an offence by himself or any person under his control S139(9).

This second defence is only available (without leave of the court) if the defence has served a notice on the prosecution, at least seven days before the hearing, giving information which identifies or assists in the identification of that other person: Barnet LBC v S &W Transport (1975), (88);

'another person' who is identified may be prosecuted and convicted of an offence under S139 irrespective of whether or not the owner is charged with an offence, S139(5): York City Council v Poller (1976), (89); A A King v Page (1970), (74).

What if the culprit is unknown: can this defence still be used? The answer appears to be yes following PGM Building Co v Kensington and Chelsea Royal London Borough Council (1985), (114) where the divisional court held that it was not necessary to identify the person and that to establish that the offence was due to the act or default of an unidentified person was a good defence.

Nothing in this section creates any liability on the part of the local highway authority for any loss or damage which might result from the granting of a permission to deposit a 'skip' on the highway S139(10).

Other cases related to 'skips' on highways are:
Wills v T F Martin (Roof Contractors Ltd) (1972), (91); Derrick v Cornhill (1970), (92); Gabriel v Enfield BC (1971), (93); Hales Containers Ltd v Ealing LBC (1972), (94).

(ii) Removal of Builders' Skips

It might be the case that following the actual depositing of a skip on the highway under a permission granted in accordance with S139 that the authorities (in this case highway or police) may require the removal or repositioning of the skip S140(2), failure to do so may bring a penalty of £50; this power exists notwithstanding that the skip was deposited lawfully under a S139 permission. It is a power which, presumably, would be used sparingly but might come about when the authorities (if police 'a constable in uniform') realise that the result of the lawful permission is that the skip is dangerous to highway users. It must be emphasized that the section does not refer to dangerous positioning but simply says that 'the highway (authority) or a constable in uniform may require the owner of the skip to remove or reposition (it)' Sl40(2). I think the assumption can safely be made that the use of this power should be 'reasonable in the circumstances'.

Should the owner fail to remove the skip the authorities may do it themselves and then may claim from the owner any expense incurred and if there is no response from the owner sell the skip and its contents, any surplus over and above the expenses being deposited in the police fund. The owner of the skip is not guilty of an offence under Sl39(4) if as a result of repositioning he fails to secure that a condition relating to the siting of a skip is complied with Sl40(9).

(iii) Approved Owners
To avoid dealing with each application individually, local highway authorities often have lists of 'approved owners' (plant or skip hirers) who have signed agreements with the authorities that they will fulfil the conditions imposed.

Hoardings

The construction and erection of hoardings is controlled by Sections l72 and l73 respectively. Before any building work is started in a street or court which will erect, alter or repair the outside of a building or take one down there will be constructed a close boarded hoarding or fence to the satisfaction of the appropriate authority, ie in most cases this will be the local highway authority; the purpose of the hoarding is to separate the building work from the street or court. This will be so in every instance unless the authorities dispense with this requirement as they would probably do regarding small works or works away from urban areas on sites where few if any members of the general public pass. The builder who erects the hoarding must if the authority requires it (Sl72):

(i) make a convenient covered platform and handrail to serve as a footway for pedestrians outside the hoarding;
(ii) maintain the hoarding and any such platform and handrail to the satisfaction of and for such time as the authority may require;

(iii) provide sufficient light for the hoarding, platform and handrail during the hours of darkness as defined in S329(l);
(iv) remove the hoarding when required by the council.

Should an application for permission to erect a hoarding be denied, appeal lies to the magistrates court and should, following the granting of permission, the builder contravene any of the above provisions he is guilty of an offence and liable to a maximum fine of £100 and one of £2 a day for a continuing offence.

Also every hoarding or similar structure which is in or adjoins a street must be securely fixed to the satisfaction of the authority (S173); if this is breached then a maximum fine of £25 and £1 a day may be imposed. Harper v Haden (50), is the leading case regarding hoardings and discusses the civil liability that may arise out of their use.

Scaffolding

Unless there is granted legal authority to do so the erection of scaffolding on the highway is an obstruction and as such is a criminal offence under both common law and statute.

Contractors (the Act refers to 'a person' making application for permission, in practice it is most likely to be the contractor rather than the owner or occupier ie the employer particularly if the work is on the basis of JCT '80 where under cl 6.1.1 the contractor has the responsibility of obtaining all statutory permissions) who wish to erect scaffolding on the highway must apply for a licence to do so from the local highway authority, who, provided that the submitted details and design of the scaffolding are satisfactory, must grant the licence. Only if the authority had reasonable grounds for stating the structure would cause an unreasonable obstruction and that an alternative structure would be less so may the authority refuse the licence.

Should the licence be refused the contractor can appeal to the magistrates court which can reverse the decision of the authority and may do the same regarding any conditions which have been imposed on a licence that has been granted.

In addition to any conditions which might be imposed by the authority S169 imposes three conditions of its own when the circumstances merit it:

(i) that the structure be lighted half an hour before sunset and half an hour before sunrise;

(ii) that any instructions or conditions relating to the erection and maintenance of traffic signs on or about the scaffold are adhered to;

(iii) to construct or to do such things to the scaffold as any statutory undertaker (ie gas, water, electricity), reasonably requests for the purpose of protecting or giving access to any apparatus belonging to or used by or maintained by them.

Failure to comply with the conditions of the licence or the statute as described above is a criminal offence with a maximum fine of £400.

Section 169 permits the erection and retention on the highway of scaffolding or other structures within the terms of the licence issued by the local highway authority but is not the only legal control as this is also exercised under the Health and Safety at Work etc Act 1974 Part 1 and the Construction (Working Places) Regulations 1966 (SI 1966 No. 94).

Local authorities have been reminded in a DOE Circular (9/77 WO Circular 8/77) not to impose conditions on a licence which either conflict with or duplicate existing legal requirements, also it must be remembered that the erection and dismantling of scaffolding is a building operation under S168 of the '80 Act and an accident, on or near a street which endangers people using the street, in the course of carrying out the work gives rise to criminal liability. The extent of scaffolding and hoardings and any precautions to be taken are usually finalised at a site meeting of all the interested parties.

'Lines'

Legal protection of highways from the encroachment of building development and also for the preservation of land possibly required, at some time in the future, for road widening, is obtained by the legal device of the local highway authority prescribing a 'line' beyond which any development work requires permission from the authority and to which conditions may be attached. The 'line' is shown on a map of the affected street and after following a laid down procedure of consultation with the notification of the affected parties eg owners, occupiers and district councils, a plan showing the line is deposited at the local land charges office in Part 4 of the register and becomes an incumberance on the land affected. Under the 1980 Act there are two types of line:

(i) a building line
(ii) an improvement line.

(i) Building Line

Under S74, '80 Act a building line is a frontage line beyond which buildings may not project without permission of the authority, the only exception being the building of boundary walls and fences. Once a line has been prescribed no new building may be erected or permanent excavation made nearer to the centre line of the highway without consent of the authority which may also impose conditions such as it considers expedient; these conditions are registerable as a local land charge and are binding on the succession in title, any breach of them is an offence with a maximum fine of £25 and £2 a day for every day the offence continues after conviction.

Should an authority refuse consent to build in front of a building a right of appeal lies to the Minister on the basis of the authority unreasonably withholding its consent.

Should the value of a property be 'injuriously affected' by prescribing a line a claim for compensation can be made against the local highway authority and in the event of a dispute arising out of such a claim it is determined by the Lands Tribunal. If the authority considers that the line is no longer necessary it may by resolution revoke it. The restrictions which follow the prescribing of a building line do not apply to statutory undertakers.

(ii) Improvement Line

An improvement line enables an authority to prevent development on land required for road widening. This is where the authority is of the opinion that a street which is a highway maintainable at public expense is narrow, inconvenient or has an irregular boundary and that it is desirable that it should be improved by widening; the authority can prescribe a line referred to as an improvement line (S73 '80 Act) on either one or both sides of the street or within 15 yards from any corner.

The restraints and conditions which are imposed following the prescribing of an improvement line are for all practicable purposes the same as those for a building line. The main difference between the two types of line is in the appeals procedure allowed for challenging a decision. Regarding improvement lines an appeal can be made against the prescribing of the line as well as for compensation for 'injurious affection' and the unreasonable withholding of consent to developing; these appeals being made to the Crown Court rather than as in the case of building lines to the Minister or Lands Tribunal.

A local planning authority is entitled to refuse planning permission for development on land for which there is a proposal for road widening and it is not acting ultra vires even when it acts to save money from the authority whose powers to prescribe an improvement or building line result in the landowner being immediately entitled to compensation for injurious affecting of his lands Westminster Bank v Beverley BC (1961), (95).

Where an improvement line is prescribed ' no new building will be erected.' '(S73(2), within the definition of buildings (S73(13)' is included'any erection'. What constitutes 'a new erection' was considered in Sittingbourne UDC v Liptons Ltd (1931), (96).

Section 4

Making up New Streets

Who should be responsible for 'making up' and maintaining new roads (in law 'private streets') has often been the source of controversy (usually related to: who pays and how much?) when following the enactment of the Highways Act 1835 only highways of a certain specified construction standard would be 'adopted' as 'highways maintainable at public expense' whilst for others no one was responsible and they became 'private streets'. In more modern times this has had the result that some roads (many of those on private housing estates built between the wars are a good example), were not made up for literally years and that those that were often fell into disrepair or were substandard and were never adopted by the local authority.

This problem of either 'unmade up' roads or ones built to unsatisfactory standards was recognised by the introduction of the private street works codes under both the Public Health Act 1875 and the Private Street Works Act 1892. These codes introduced the concepts of 'new streets', 'construction costs shared on a frontage basis' and even an element of democracy, whereby, in certain circumstances a majority of the frontages concerned could insist (1875 code only) on the road being made up to the specification and standards of the local bye-laws under the supervision of the local authority, thereby being suitable for adoption as a highway maintainable at public expense.

With the enactment of the Local Government Act 1972,the code of 1875 was repealed and more recently the code of 1892 has been re-enacted in sections 205-218 of the 1980 Act, making it the generally applicable private street works code (unless there is an applicable local Act) rather than the optional code it had been under the adoptive Act of 1892.

The procedures by which private streets become highways maintainable at public expense and other powers related to the construction and financing of street works are set down in sections 186-237 of the 1980 Act; included is the power to make bye-laws which control the standards to which roads will be built (S 186); these roads are referred to as 'new streets'.

'New Streets'

What exactly is a new street is not defined or clearly stated in the statute; to obtain clarification it is necessary to refer to case law, itself not always too helpful as the courts insist on dealing with each case as a matter of fact which makes the development of a rule or principle of law difficult.

New streets can be newly laid out roads such as new estate roads (dealt with usually under the 'Advance Payments Code' (p.51) or by 'Agreement under S38' of the 1980 Act p.53) or they can be existing highways (maintainable at public expense or not, ie private streets) as with 'infill sites', or where existing roads are extended by building on one or both sides of the street, authority for which lies in S187, 1980 Act. If a developer builds on a plot of land and the houses front on to what must become a street ' in the popular sense of the term... or as is commonly understood' then clearly a street is being laid out and the byelaws for new street construction must be complied with.

A highway may be a street as defined in the Act (S329, 'street includes any highway and any road, lane or footpath, square, court, alley or passage, whether a thoroughfare or not and includes any part of a street') and by building on to this 'street' it becomes a 'new street' for the purpose of the byelaws. This rule comes from Robinson v The Local Board for the District of Barton-Eccles, Winton and Monton (1883), (97); where it was held that the words 'new streets' are not confined to streets being built for the first time but also applied to an old highway, formerly a country lane, which had been a 'street' within the meaning of the Public Health Act 1875 (as above), and which by building houses on either side of it had recently become a street in the popular sense of the term.

The implication of this seems to be that if buildings are erected abutting a highway which is not a street as 'commonly understood' but which as a result of building activity becomes so; then the highway becomes a 'new street' and its construction must comply with the new street byelaws. It also seems to envisage that the construction will be of a road as 'understood in the popular sense' Devonport Corporation v Tozer (1902), (98).

The term 'new street' was again at the centre of judicial interpretation in Astor v Fulham (1963), (99); where the nature of a street was changed making it into a 'new street' even though it must have been a street for some years and had been a highway since 'time immemorial'.

Local authorities pass byelaws (S186 1980 Act) to regulate the level, the width and construction of new streets in their area including provision for sewerage and if necessary separate sewers for foul and surface water requirements. Byelaws usually also contain requirements relating to the deposit of plans (with their possibility of revocation if work has not commenced within three years) and the subsequent inspection and testing of the sewers prior to adoption.

When development takes place along the side of an existing highway, maintainable at public expense or not, the local authority may declare (S188) its conversion into a 'new street' and 'prescribe the centre line of the new street and outer lines defining the minimum width of the new street... required by byelaw.' When this order is made it is registerable in Part IV of the local land charges register.

The effect of the order is to limit a landowner's use of his land in that:

(a) no building (including a wall) may be erected on land between the two outer lines;
(b) when building work commences on adjoining land the land between the outer prescribed lines becomes part of the highway and the owner must remove any boundary fences, walls etc and bring the level into conformity with that of the existing highway.

Notice of the prescribing order must be given, which allows appeal to the Crown Court.

The strips of land which lie between the existing highway and the prescribed lines remain in the ownership of the landowner who can exercise the rights of ownership with the exception of building on the land. When the street is made up it is done so under the appropriate private street works code, the work on the 'strips' is paid for by the respective frontages.

Three cases give good examples of some of the problems associated with the concept of frontages: Buckinghamshire CC v Trigg (1963), (100); shows that the owner of a ground floor flat and front garden is liable for apportionment and that the upper flat, even though having access to the highway across the front garden, does not; the upper flat does not adjoin the street.

In Warwickshire CC v Adkins (1968), (101); the dispute was between houseowners and the developer who had retained ownership in a strip of land (12 feet wide) which as a condition of the planning permission remained at the side of the existing road undeveloped to enable the road to be widened in the future. The houseowners were apportioned the costs of the streetage; they objected claiming the landowner (the developer) was liable. On appeal it was held that the house was 'premises fronting the street' and that lack of evidence of dedication by the developer was irrelevant.

Again the arguments in Ware v Gaunt (1960), (102); are regarding frontages and who pays. Footpaths were excluded from the adoption procedures following the 1835 Act being considered automatically maintainable at public expense. In this case in 1885 a road was laid out but only ever used as a footpath; it was held that it was a new street and the frontages had to pay.

Where a 'new street' will form a main thoroughfare or the continuation of one, or a link between two of them, the authority can insist (S193) on a width of street as it determines and this width may be imposed as a condition on the passing of the plans. If the width required by the authority is in excess of 20 feet more than the normal maximum (stated in byelaw) then the authority should pay compensation; any dispute being settled by the Lands Tribunal. Frontages would bear the cost of 'making up' but this should not be more than that which would have been incurred had the street been the normal maximum width; any dispute over this is determined in the magistrates court. Any person aggrieved by a condition imposed by this section may appeal to the Crown Court. (This section provides for three different tribunals to determine disputes which may arise from its provisions, albeit in differing matters ie Lands Tribunal, Magistrates Court and the Crown Court). Similar provisions exist (S196) to allow the widening of an existing street on one side only.

Byelaws which are normally strictly enforced may be relaxed where the authority consider that it 'would be unreasonable in relation to a particular case... they may with the consent of the Secretary of State relax the requirements of the byelaws...' the

Secretary of State shall take into consideration any objection...
received by him', (S190(3).

Town and Country Planning

To achieve a measure of uniformity in the application of planning
control in relation to new streets the General Development Order
1977 (SI 1977 No 289) provides that where planning permission is
granted for development which consists of or includes the building
or laying out of a new street none of the bye-laws relating to new
streets (save those requiring separate drainage) shall apply to that
development.

In modern practice the use of new street orders and bye-laws may
cause difficulty, as the standards laid down in the bye-laws are
sometimes more stringent than circumstances really require,
especially in the case of small housing estates. Consequently some
planning authorities have obtained development orders from the
Ministry of Housing and Local Government under S24 of the 1971
Act, relaxing generally the byelaws provided that the schedule of
minimum street widths, which is included in the order, is complied
with.

Enforcement

As a natural concomitant of the previous provisions the 1980 Act
gives power to the authorities to enforce any conditions imposed
and also to require removal or alteration of any work which does
not conform to those conditions; fines may also be levied 'on a
person who executes work ... otherwise than in accordance with [a]
condition' S198; conditions are enforced against the owner of the
land to which they relate.

There are three separate and distinct procedures any one of which
may be followed and will ensure the construction of highways to a
standard which will enable them to be adopted by the local
highway authority as 'highways maintainable at public expense'
should they wish to do so or if the frontages insist on it (S204);
adoption would be the objective of the majority of builders and
developers:

(i)   Private Street Works Code (S205-218) 1980 Act
(ii) Advanced Payments Code (S219-225) 1980 Act
(iii) Agreement made under S38 1980 Act.

Private Street Works Code
This is invoked when an authority is dissatisfied with the condition of a private street and it ensures that it is 'sewered, levelled, paved, metalled, flagged, channelled, made good and lighted' and in accordance with the code, the costs incurred are shared between the frontages; the work, in a majority of cases, being carried out by the authority itself. The authority can include any works it thinks necessary for bringing the street into conformity with other streets whether maintained publicly or not, including building separate sewers for sewage and surface water.

Following the passing of a resolution to carry out the works 'the proper officer of the council' (highway engineer) shall prepare specifications and drawings of the proposed work with estimates of probable expenditure and with a provisional apportionment of that expenditure between the frontages involved, all of which are prepared on the basis of paragraphs 1 to 4 of Schedule 16 (S205).

This whole procedure is set in motion when the local authority passes a resolution to execute the works. This is then confirmed by a 'resolution of approval' after the submission to the authority of the completed specifications and estimates, details of which may be amended as the authority thinks fit. This is though limited to amendments involving less work than previously identified or if not so then it certainly cannot include any extra work. This is based on the proposition that the original resolution indicates the authority's dissatisfaction with certain existing works which should cover every area of work which the authority wishes to carry out, the authority will not be allowed to come along and make an amendment to improve the street lighting if it had not been included in the original resolution, Ware UDC v Gaunt (1960), (102). The efficacy of this code depends on the attitude of the authority as it is the authority which will invoke it or not; in less aware authorities private streets may exist for years without being 'made up' and adopted (see p.11) due to lack of action on the part of the authority concerned, possibly due in part to the likely increase in road maintenance costs adoption would eventually incur and which could be in direct conflict with other financial policies of the authority.

Whilst nothing much could be done about existing private streets, (it tends to be a diminishing problem as they are either made up or the older ones demolished) something could be done regarding new ones and the procedures introduced were designed to ensure that roads were built simultaneously with the construction of the houses or other buildings being erected; thereby ensuring streets of an 'adoptable' standard from the beginning. It is with this end in mind that the remaining two procedures were introduced.

Advance Payments Code
The purpose of this code is to secure the payment of the expenses incurred in the execution of private street works which are adjacent to new building work. The code applies in most parts of the country but is an adoptive measure carried out under Schedule 15, 1980 Act.

The code applies where it is proposed to erect a building for which plans under the building regulations are required to be submitted to the local authority and that the proposed development fronts on to a private street in which the 'streets authority' has power under the private street works code to require works to be carried out on the land unless the owner (or previous one) has paid or secured such sum as is required in respect of the street works as determined by and to the satisfaction of the authority. It is for the authority to decide what form the security shall take, but, under S76 of the Building Societies Act 1960, it seems clear that a mortgage or charge on the land is intended and that it is not to have the interpretation of a prior mortgage for the purpose of a building society advance (S76, Sch 6, 1960 Act). Failure to comply with these requirements is an offence carrying a maximum fine of £100; should a person, not the owner, for example, the contractor, be charged it is a good defence to show he had reasonable grounds for believing that the owner had deposited or secured the money required; only the streets authority can commence proceedings.

(i) Exemptions to the Code
There are exemptions to the general rule regarding the deposit or securing of funds for making up roads S219 (4), given the relevant circumstances the exemption is automatic whereas in others it is a discretionary power of the authority.

It can be of great value for the architect or builder to approach the authority, at the time say of submitting plans for planning or regulation approval, to discuss (or enquire of) the authority's application of its discretionary powers viz advance payments for making up roads and establish (or even persuade!) the authority's position regarding the site in question; to obtain exemption is a benefit to the client or developer and may even be influential as to whether or not a project continues.

The following are the exemptions in S219 (4):

(i)   where the private street works code exempts from liability the owner of land; under S215 places of public worship are exempt with the authority bearing the cost of the street works;

(ii)   where the building is to be erected within the curtilage of an existing building (no definition of curtilage is given) Sinclair-Lockhart's Trustees v Central Land Board [1951]; (104);

(iii)   where plans were deposited before the code was adopted;

(iv)   where a S38 agreement exists (see below);

(v)   where it is unlikely that the whole of the street will be built up within a reasonable time, eg isolated development;

(vi)   where the proposed work is on a street which does not join or 'run into' a highway maintainable at public expense;

(vii)   where the street is less than 100 yards long and is at least 50 per cent built up, inclusive of both sides;

(viii)   where the development is on a street which is substantially built up the authority may exempt by notice the proposed building from this section (distinguished from iii);

(ix)   where the land is in the ownership of certain nationalised corporations or local authorities, eg British Rail;

(x)   where the development is industrial and the building work is carried out by a company the objects of which allow the provision of buildings for other than company use and the constitution of which prohibits the distribution of the company's profits to its members and also where the cost of building is to be defrayed by a government department;

(xi)   where the development is largely industrial and it is unlikely that the powers of the 'streets authority' will be exercised within a reasonable time; this exemption is by a resolution relating specifically to the particular development and passed by the authority.

Money deposited prior to the passing of this resolution has to be returned or securities released and if the land concerned has since been divided, repayment is apportioned accordingly to frontages.

(ii) Amount of Payment

Within one week (S220) of a developer submitting drawings for planning and building regulation permission for a proposed building on a private street, the district council must inform the street works authority which in turn must, within a further six weeks notify the person by or on whose behalf the drawings were submitted of the amount of money to be deposited or secured which in the authority's opinion would be recoverable under the street works code, notice is served on the person depositing the drawings, payment when due is from the owner. In Whitstable UDC v Campbell (1959), (105); it was held that a notice sent to the office of the client's architect was sufficiently served.

In any dispute over that amount there is a right of appeal to the Minister. Simple interest is payable on the amount deposited at a rate fixed by the Treasury under S2 Public Works Loans Act 1964.

If the street is made up by someone other than the authority a refund is due to the person at whose expense the work was carried out, frontages are however entitled to make representations before any money is repaid or security released (S221).

In the event of the amount being in excess of that required, the authority will refund the difference to the owner, for the time being. If ownership has changed in the meantime a prudent owner may insert a clause in the contract of sale to the effect that any refund is passed on to him; this may well be a wise thing to do as many authorities err on the right side when preparing the estimates (S222). Similar provision exists for refunding monies should a project be abandoned or is not commenced within the three years following the granting of permission. If having abandoned a project the owner wishes to proceed (within three years) the notice referred to above (S220) must be served within one month rather than six weeks.

(iii) Power to Insist on Making up Road
It is within the authority's discretion as to whether the private street works codes are applied but owners can enforce the making up of streets where at least one of the owners has paid a deposit or given security under the Advance Payments Code (S229).

Agreement under S38

The major objection that a builder or developer may have to the Advance Payments Code is that money is being paid out before any work is done, therefore capital is tied up and even if simple interest is paid the process creates an extra cost to be borne by the builder and consequently the building owner, usually a house owner.

An alternative method of financing the building of new estate roads is provided by S38 of the 1980 Act whereby the builder agrees with the authority that the roads will be built to the authority's specifications and under its direct supervision. On completion the roads are dedicated by the builder as highways and the authority adopts them as highways maintainable at public expense.

What is of real importance to the contractor or developer is that he does not have to pay in advance (ie deposit or security (S219 (4) (d)) as in the Advance Payments Code, there is no financial outlay other than for the work as it progresses.

Presumably as a result of the historically high building industry liquidations and bankruptcies, authorities commonly require agreements of this nature to be supported by a bond for due performance. The respective liabilities of the builder, the local authority and the security are considered in National Employers' Mutual General Insurance Association v Herne Bay UDC (1972), (106); which arose following the voluntary liquidation of the builder.

Private Roads
The distinction has already been made (1) between highways and private roads; essentially it being that private roads remain in the control of the landowner thereby denying the public access 'as of right' which would create a 'public way' or highway.

Private roads may be a single street, a private estate (in the true sense of the phrase) of streets, a passage, or courtyard etc and there are plenty of examples in both town and country identified by the sign 'private road'; an excellent illustration being the private estate at the rear of Nottingham Castle, where to walk or drive on the streets without the intention of calling at one of the houses (implied term for right of access) is a trespass to land. Sometimes these private roads may be described as occupation roads which were (are) exclusive to the people who occupy the street; as time passes there are now many 'Occupation Roads' which have become highways maintainable at public expense.

(1) p.2, 3, 5, 12

Private land can also be subject to public rights where sites are not developed up to the boundary of the plot but fall short, with land from the site being at a level of and becoming contiguous with the footpath on the highway. To retain ownership and control (the landowner may wish to develop up to the boundary at some future time), it is often the practice to insert steel studs in the footpath in line with the boundary and a notice or plaque fixed to a wall or even inserted in the footpath itself, claiming 'the land within the studs is private land and not intended to form part of the highway' or some such statement.

Adoption of Streets

Adoption comes about in one of three ways:

(i)   at the discretion of the authority
(ii)   by some 'democratic' action, eg majority of frontages or
(iii) by agreement under S38 (p.53).

(i) At the Discretion of the Authority

Where street works have been carried out in a private street, the authority may adopt the street, this being brought about by simply posting a notice to that effect in a prominent position in the street, and in one month's time from the date of the notice, so it becomes. This does not occur if the owner, or if more than one, a majority in numbers, notify their objections to the authority within that month, this in turn may also be overruled, if the authority appeals to the magistrates court within a further two months (S223 (1) (2)).

Should the authority not wish to adopt after the work has been satisfactorily completed, a majority of rateable value of owners can insist on it and following an application to the authority of that nature the authority shall display a notice in a prominent position in the street, declaring that: the street is a highway maintainable at public expense and 'thereupon the street shall become such a highway.'

(ii) 'Democratic' Action

When street works have been completed under the Advance Payments Code or where a minimum of one instalment has been paid or secured, and the works have not been carried out a majority of owners in numbers or as many of those owners who own between them more than half the aggregate length of the frontages can insist by notice to the authority that the street is adopted or made up and adopted as the case may be (S229).

(iii) Agreement under S38
If the works are carried out to the satisfaction of the authority, adoption is automatic.

Serving of Notice
On a number of occasions when applying sections in the Highways Act 1980 it becomes necessary for the authorities to serve notices on landowners, developers and such like which must be in accordance with S322. This section is fulfilled if the notice is served on the architect (and presumably other agents and advisers, surveyors etc) acting for the owner or occupier of premises: Whitstable UDC v Campbell (1959), (105).

Section 5

Civil Liability

Someone suffering personal injury or property damage following an accident as a result of building work or associated activities either on or near to the highway may find it possible to seek a legal remedy for the injuries sustained, against one or more of the following:

(i)   the building owner, ie the client; may be the owner of the site or just the occupier with the obligation to repair;
(ii)   the contractor or sub-contractor;
(iii)   the construction professional, eg the architect, engineer or surveyor;
(iv)   the Local Highway Authority.

The Building Owner

Under English Law it is normal that individuals (included are companies) are liable for their own legal shortcomings; only in certain clearly defined circumstances does law make one person liable for the wrongful act of another, ie impose the principle of vicarious liability; work carried out on or near to the highway is one such instance.

The liability is imposed on the employer (as occupier) for the general repair of the building in relation to members of the general public using the adjacent highway; this liability is strict and is therefore applied without negligence on the part of the employer. It is applied when personal injury or property damage is caused by the contractor to passers-by or their property. This is established by Tarry v Ashton (1) where a lamp fixed negligently by an independent contractor fell off the building and injured Tarry, a passer-by; irrespective of any negligence on his part, Ashton was liable. This decision was made by a two to one majority in favour of strict liability existing in these kinds of circumstances. Blackburn J. dissented arguing that fault should exist. Notwithstanding this decision, there has been judicial opinion contradicting it, as expressed by Williams L.J. in Barker v Herbert [1911], (63); where he said 'there can be no liability...of the possessor of land in such cases unless... he himself ...created that danger which constitutes a nuisance to the highway, or that he neglected for an undue time after he became, or, if he had used reasonable care, ought to have become, aware of it, to abate or prevent the danger or nuisance.' His decision appears to introduce the concept of fault into cases of this nature.

However strict liability was reiterated in Wringe v Cohen (1939), (64), where the Court of Appeal stated: 'If owing to want to repair, premises on a highway become dangerous and, therefore, a nuisance and a passerby or an adjoining property owner suffers damage by their collapse, the occupier, or owner if he has undertaken the duty of repair, is answerable whether he knew, or ought to have known, of the danger or not.' The court did make two exceptions: (i) 'where the dangerous circumstances are created by a trespasser', and (ii) where there is 'a secret and unobservable operation of nature, such as subsidence under or near to foundations of the premises.' This latter point, arising out of Wringe which dealt with a building collapse on to private land and not the highway, is considered by some to be rather odd.

In both Tarry v Ashton and Wringe v Cohen the building owner is liable irrespective of his own negligence for the actions of the independent contractor. The only way the building owner can avoid this strict liability is by being able to distinguish any incident from these two leading cases. The emphasis is placed on what is called collateral negligence, ie that the action complained of was collateral to the main work which was commissioned by the employer, as shown in Padbury v Holliday and Greenwood (60) and Reedie v L & N W Railway Co. (61).

Obviously these principles have an influence on the builder's work in assisting to determine actions for which the contractor would be liable and those which are the responsibility of the client. An illustration of this might be gutters which fall off a building and injure a passer-by, this would be the liability of the building owner whereas the collapse of the scaffolding erected to repair or replace the gutters would be collateral and therefore the liability of the builder.

If work is carried out near to the highway and injury is caused to highway users the building owner may be liable only if the work was inherently dangerous in Salsbury v Woodland and others

[1970], (62); the owner was not vicariously liable for the negligence of his independent contractor when trees which were felled negligently fell on to the highway resulting in personal injury.

Another exception is where the problem arises from natural sources; the concept of latent defects is acceptable and negligence has to be established; this was so in both Noble v Harrison [1926], (43); and Caminer v Northern & London Investment Trust (1951), (33) where trees which, unknown to the owners, were diseased, collapsed causing damage to passers-by.

In these circumstances there is no involvement of an independent contractor: the liability or not relates to the land owner.

A building owner whilst being unable to avoid the legal liability if it is warranted may well escape the financial implications by the insertion of an indemnity clause in the contract; this, as referred to on p.32, may have its limitations but is a good example of transferring, not the legal liability, but its financial consequences. A similar clause (cl. 22) is in the I.C.E. standard form of contract.

The Contractor and the Sub-Contractor

Whenever a contractor does work on or over the highway he owes a duty of care to the other users of the highway and is liable for any personal injuries or property damage that may arise from a breach of that duty. To fulfill this duty the contractor must in carrying out his work minimise the likelihood of accidents occurring by instigating appropriate protection methods, for example lighting works at night and the existence of warning notices drawing the attention of the public to possible dangers.

A contractor may face legal action from three possible sources: firstly he may face prosecution by the authorities for breach of either the common law (public nuisance) or statute such as the Highways Act 1980, which on conviction results in a fine or possibly the granting of an injunction; secondly a civil action for damages might also follow such a breach where special damage can be established. Thirdly, the action may be based on negligence where personal injury and property damage has been sustained; the claim is for money damages.

The very nature of a contractor's work on the highway will result in an obstruction but as referred to previously authority can be obtained making these construction activities lawful.

It is established that the contractor owes a duty of care to other users of the highway and this applies equally to the disabled as to the able bodied. In Haley v London Electricity Board (1965), (66) the duty was held to be owed to a blind person who was injured by falling over some builder's rubble 'it was reasonably foreseeable that a blind person might pass along the pavement...' per L.Guest, he also stated that 'a contractor must have regard to all road users which include blind and other persons'. The question of 'other persons' here raises the point that if the *ejusdem generis* rule is applied it presumably means that other disabled people would pass by, say, in wheelchairs, with the implication that the contractor should make provision for them in terms of substitute footpaths etc.

In carrying out work of this nature the contractor must comply with any conditions attached to a statutory permission (1) also the works must be carried out in a reasonable manner *vis-à-vis* the general public.

What is 'reasonable' is a decision of the court depending on the circumstances but it is safe to assume that a higher standard would be required in a busy urban centre than say in a scarcely populated rural district. To carry out work to the standards set out in any conditions attached to a permission would, more than likely, reach the standard of 'reasonable in the circumstances' for that particular job. Where conditions imposed are basic the contractor in order to fulfil the duty may have to impose a higher standard upon himself: it is not axiomatic that to fulfil conditions imposed by the local authority will also meet the common law standard necessary to fulfil the duty of care. The contractor must be aware of this possible difference and be constantly alert to it, particularly during the project planning process as it may have financial implications.

Examples of work which can be strictly controlled are hoardings, scaffolding, parking of skips where specifications can include the provision of temporary footpaths, handrails, warning notices, lighting etc. These can be defined quite accurately but excavation work can be more difficult and problematical but nevertheless the public does require protection from it. Each individual job should be considered on its merits; automatic assumptions should not be made.

Certainly the standard imposed by the courts is high as is demonstrated in Pitman v Southern Electricity Board (1978), (90);

(1) Ref. p.38, 41, 42

where at the end of the working day a steel plate was placed over a small excavation (850 x 850mm) in the footpath. The plate was 3mm thick. Later 'Mrs P' a 78 year old woman tripped over the plate and broke her wrist. The electricity board were found negligent for leaving the footpath uneven. The Haley case referred to above (66) also deals with excavations and the provision of barriers.

Reference has previously been made to the possibility of the employer being held vicariously liable for the negligence of the contractor: exactly the same principle can be applied to the main contractor regarding the conduct of sub-contractors. In Holliday v National Telephone Co (1899), (68), the company were liable for the negligence of their independent contractor, a plumber, who was soldering joints of a cable, which splashed a passer-by on the highway. The company were liable because 'they were under a duty to take care that those who do work on their behalf do not negligently cause injury to persons using the highway'. In 'doing work on their behalf... and owe a duty to persons using the highway', the arguments relating to collateral negligence may be available as illustrated in Padbury v Holliday and Greenwood (60).

The Construction Professional

The professional assisting and advising the client may possibly attract a liability whilst fulfilling these obligations; this is most apparent in the offering of advice.

The professional owes a duty of care to the employer when preparing the design work and offering advice related to it. The duty 'to act diligently' is the legal basis on which the obligation to pay professional fees is based. The courts expect quite a detailed knowledge on the part of the construction professional; this was highlighted in B L Holdings v Robert Wood & Partners (1979), (116) where an architect was held liable in the court of first instance in negligence for not advising the client that the local planning authority were wrong in a decision that was made which excluded car parking and the caretaker's penthouse flat from the 10,000 sq ft office development certificate. B L Holdings lessee withdrew when they discovered a certificate has not been issued, and B L sued the architect. The Court of Appeal thought that it put the duty too high and reversed the decision. It does, though, still highlight the necessity of the professional having a good standard of legal knowledge appertaining to any particular discipline.

It was always thought that this duty was based on the contract between the professional and the client but this was widened to include third parties by such cases as Hedley Byrne v Heller & Partners (1963), (117) and Clay v Crump (1964), (118). The liability associated with giving an opinion may also be a breach of S3 of the Health & Safety at Work etc Act 1974, where the Act imposes a duty on employers and the self-employed, to persons other than their employees. This could widen the scope of liability to the men on site or members of the general public. Breach under the Act is a criminal offence and may result in a fine being payable rather than damages.

These examples coupled with the approach of the judiciary to highway cases give a good indication of the possible seriousness for the professional should things go wrong.

The Local Highway Authority

The liability of a local highway authority at common law for personal injury and property damage which results from the condition of the highway is founded on the highway being publicly maintained and the application of the doctrine of misfeasance(1). For many years the public authority was not liable for damage caused by a lack of repair or maintenance, Russell v Men of Devon (1788) (19); only in the event of damage following misfeasance, i.e.carrying out repairs in a negligent manner, would liability lie with the authority. This defence existed until it was abrogated by S1 of the Highways (Miscellaneous Provisions) Act 1961 which has now been superseded by S58 of the '80 Act. An example is illustrated in Burton v West Suffolk CC(1960), (113).

Local highway authorities now have a duty to maintain (includes repair S329 (1)) highways maintainable at public expense (S41 '80 Act), and failure to do so which results in damage to a lawful user of the highway can be the basis of an action claiming damages based on a breach of statutory duty.

Under S58 a local highway authority has the defence available that it has 'taken such care as in all the circumstances was reasonably required to secure that the part of the highway to which the action relates was not dangerous to traffic'. For the purposes of this defence (S58 (2)) the court shall have regard to the following matters:

(a) the character of the highway, and the traffic likely to use it;
(b) the standard of maintenance appropriate for a highway of that character and used by such traffic;

(c) the state of repair in which a reasonable person would have expected to find the highway;

(d) whether the authority knew, or could reasonably have been expected to know, that the condition of the highway was likely to cause danger to users;

(e) where the authority could not reasonably have repaired the highway before the action arose, what warning notices had been displayed; for example 'beware potholes' or when carrying out maintenance work 'raised manholes' or 'ramp ahead'. Consideration of the standard to which highways are repaired and maintained is dependent upon judicial interpretation, i.e. the rules of common law; but as can be seen from S58 what is 'reasonable' will be important.

There was a similar outcome in Whitaker v West Yorkshire Metropolitan County Council and Metropolitan Borough of Calderdale (1981), (115) when 'W' was injured as a result of slipping on an icy patch on a roadway. 'W' claimed that the authority were in breach of their statutory duty to maintain the highway and that the defendants had negligently allowed water to escape onto the highway from land which they owned. 'W' lost her case; she had been unable to prove she had slipped on ice formed from water coming from any land.

It was suggested, obiter, that the authority would in any case have been successful with the defence set out in (d) above.

In Littler v Liverpool Corporation (1968), (103), Cumming-Bruce J. asks: 'what standard is a highway authority under a duty to maintain?... where the course of action of the plaintiff suffered personal injury by reason of the failure to maintain the highway, the plaintiff must make out a case that the highway was not reasonably safe, that it was dangerous to the relevant traffic... What then is the test of "dangerous" in this context... The approach foreshadowed obiter in the speeches of the Court of Appeal in Griffiths v Liverpool Corporation(1968), (107),... the test in relation to a length of pavement is reasonable foreseeability of danger. A length of pavement is only dangerous if, in the ordinary course of human affairs, danger may reasonably be anticipated from its continued use by the public who usually pass over it.' What is dangerous? This is very much a matter of fact, the particular circumstances of each incident, case, being considered on its own merits as is emphasised by Lawton J. in Rider v Rider ([1973), (108), 'In most cases proof that there were bumps or small holes in the road, or a slight unevenness on a pavement, will not amount to proof of a danger to traffic through failure to maintain. It does not follow, however, that such conditions can never be a danger to traffic. A stretch of uneven paving outside a factory probably would not be a danger to traffic but a similar stretch outside an old people's home, and much used by the inmates to the knowledge of the authority might be.'

What is dangerous seems very much to be based on the facts of the case as they are explained: so that in Meggs v Liverpool Corp (1968), (112) the plaintiff lost her case because users of the highway must expect some unevenness and the evidence was insufficient to show the pavement was dangerous or that the obligation to repair had been breached. The contrary was the decision in the Pitman case, where case for unevenness of only 3 mm was held sufficient to be dangerous and in Littler the unevenness was 12 mm (½ in) but was not dangerous. Hardakers v Idle District Council (1896), (109), shows that the authority cannot pass its liabilities off to independent contractors and Burnside and Another v Emerson and Others (1968), 741, (110), shows contributory negligence is available as a defence to the authority. In Griffiths (107) the Court of Appeal held that the duty of maintenance was strict, subject only to the statutory defences, but in Haydon v Kent C C (1978), (111) where 'H' slipped on an icy path, Goff L.J. says: 'the plaintiff must prove either that the... authority is at fault apart from merely failing to take steps to deal with the ice, or, which is the point of this case, that, having regard to the nature and importance of the way, sufficient time had elapsed to make it prima facia unreasonable for the authority to have failed to take remedial measures. Then the authority is liable unless it is able to make out the statutory defence.'

From these cases a member of the general public can study the circumstances of any particular incident and make an assessment as to whether or not they meet the likely criteria to found a successful action. But from a construction viewpoint they highlight some important procedures and responsibilities for highway engineers and other highway authority professionals.

Section 6

Short Case Studies

1   **Tarry v Ashton (1876) 1 QBD 314**

Rule nisi obtained by the defendant for an order of non suit in an action in which the plaintiff claimed damages for injuries caused by the defendant's gas lamp falling on her head.

The plaintiff was a barmaid living in Dulwich and the defendant a licensed victualler who kept the Hampshire Hog Inn in the Strand.

The plaintiff was walking along the Strand at about 3.30 pm on 15 November 1874, when opposite the defendant's house, one of the gas lamps hanging from the house over the pavement about 15 ft fell and injured her in its fall. At the time of the accident the gas was out of order and the defendant had engaged a gas fitter named Weaver to put it to rights. The gas fitter had placed a ladder against the bracket of the lamp and was just mounting the ladder for the purpose of reaching the lamp and blowing the water out of the gas pipe, when the bracket gave way from the weight of the ladder and the lamp at once fell. In the preceding August the defendant had contracted the services of a gas fitter named Chappell for the repair of all his lamps, including the one in question. Chappell had done certain repairs, for which the defendant had paid him.

2   **Ex Parte Lewis (1888) 21 QBD, Digest 2601**

The case arose out of Lewis who wished to hold a public meeting in Trafalgar Square 'as of right'. Hence the statement that the only right with regards to the highway is 'A right for all her Majesty's subjects at all seasons of the year freely and at their will to pass and repass without let or hinderance' (p.197 QBD).

3   **Goodtitle and Chester v Alker and Another**
     (1757)1 BURR 133;427;97ER 231, Vol 21, Hals. p.72

John Gotley built a house at Halfords Gate but doing so had encroached on to some of his neighbour's land. John Gotley had also built a fence at the front that also encroached on to the King's highway. Chester agreed to rent the land which Gotley had encroached upon for a sum of 6s. 8d. per annum, the lease being 100 years.

Gotley was not allowed to encroach upon the highway as the subjects of the sovereign have a right of passage over the surface of the highway.

The subjects of the sovereign have nothing beyond the right of passage over the surface of the highway. The freehold in the soil and all profits therefrom belong to the owner of the soil and he has the right to work mines under the highway, lay pipes to carry water, and so on. If any person encroaches on the highway the landlord may recover possession subject to the right of the sovereign's subjects to pass over the highway. An action for trespass will be served for any injury done to the highway.

4   **Hue v Whitely (1929) 1 Ch 440**

'H' and 'W' were neighbouring freeholders: a rough road existed between the properties, from London Road to Box Hill, regarding which 'H' took a conveyance in 1924. 'W' without the consent of 'H' opened a gate in his boundary for pedestrians. 'H' applied as freeholder for an injunction to stop 'W' trespassing. Evidence was given of public user for the purposes of pleasure.

Held: Evidence of public user led to a presumption of dedication, the motive was irrelevant.

5   **Fairey v Southampton County Council (1956) 2 QB 439**

In 1954, 'a landowner applied to Quarter Sessions for a declaration that no right of way existed over a path on his land shown as a public path on a map prepared by the local council (i.e. Southampton C C).

Quarter Sessions discovered that the path had been used by the public with no interruption from 1895 to 1931.

In 1931 the then land owner objected to the use of the path by the public other than the local residents, and since then had successfully 'turned off' such persons and by doing so had shown an intention not to dedicate the path as a highway. But a public right of way had been dedicated by Section 1 of the Rights of Way Act 1932.

The application was dismissed.

The landlord appealed to the Divisional Court which held that the right of way was brought into 'question' within the meaning of Section 1 1932 Act.

Because the landlord for the first time refused to allow the public to use a path which they had been using for 20 years or more without interruption it was not necessary to bring the case into the act.

6   **Attorney-General v Esher Linoleum Co Ltd (1901) 2 Ch 647**

Where there is a public right of footway across land and a certain amount of surface of land lying along the course of the public footway devoted to traffic, even if it is private carriage traffic, the owner of the soil must be taken as to have dedicated it to the public as a footway, so much of the surface as he has in fact devoted to traffic.

The land in question was 250 yards long, and varied in width from 55 ft at the enclosed end to 40 ft at the other. On the south was a railway separated by a ditch, hedge post and railfence. On the north was the defendants' premises, manager's residence and other buildings. On the west was the River Mole. The defendants owned the soil strip opposite their premises which was fenced off by the river but at the northwest corner there was a footbridge over the river, and a public right of way over it. From the east the strip was approached by a road through an arch under the railway line. The road then turned west towards the river. There was also a private carriageway along the strip from the arch to the mill gate. The plaintiffs contended that the carriageway extended up to the river. The defendants alleged that the public footway coming from Esher was along the southern side of the strip by the side of the railway and it crossed over the strip in a starting direction towards the mill gate and ran along the northern side by the side of the defendants' premises to the bridge. This left a triangular piece of land which the defendants had enclosed.

J. Buckley stated that if there is a public right of footway it is, of course, impossible that the defendants can enclose it so as to prevent the public from using that right.

7 Williams-Ellis v Cobb and Other (1935) 1KB 310

The action out of which this appeal arises was brought in the county court of Pwllheli before Judge Arterus Jones. The Plaintiff is the owner of an estate in Caernarvonshire, and the defendants were sued for trespass. Their defence was that they were not trespassers but acting lawfully in exercise of a public right of way. There was also a claim based on a local custom or right of the inhabitants which is now immaterial. Neither the Attorney-General nor the county council were parties, thus no question of public right could be determined by the proceedings, in which the issue of the right of way arose merely as between the parties as being the defence set up by the defendants.

A right of way may be proved even though it does not lead to a public place. A public way may, for instance, lead to the sea although the public have no right of crossing the foreshore. If there is a public way down to the sea which has been used in connection with fishing and navigation, it may be contended that the public have a further and an ancillary right to go from the end of the way over the foreshore as the tide recedes to fish or to land or embark goods, fish or passengers.

8 Lewis v Thomas (1950) 1 KB 438

Public use of a footpath of up to 40 years interrupted from time to time by locking a gate across the path was prima facia an interruption within the meaning of S1 Rights of Way Act, 1932, even if not challenged. The absence of any intention to challenge the right of the public to use the way is material to the question whether in fact there has in fact been any interruption within the meaning of S1. The gate was locked simply to prevent animals straying into the cornfields, there was no intention to stop the right of way. Farmer suing for trespass, user claiming public right of way.

Held: There had been no interruption in fact of the user since the locking of the gate had been done at such times and in such circumstances as not to be likely to interrupt, and not in fact, to have interrupted the use of the way.

9 Sandgate UDC v Kent CC (1898) 79 LT HL, All ER

The appellants Sandgate UDC made a legal claim against respondents Kent CC under Section 11 of the Local Government Act

1888 in respect of costs over previous five years ending March 25 1893 in respect of maintenance and repair and reasonable improvement of such part of main road running from Folkestone to Hythe in Kent. The part referred to was sea front and abutted the shore. The county council, Kent, repudiated the claim.

However, the county council were ordered to pay £6,188 to the urban district council. Each party were ordered to pay their own costs.

10   **Ramus v Southend Local Board (1892) 67 LT 169**

The plaintiff, Frederick Francis Ramus, claimed a mandatory injunction against Southend Local Board for the removal of a fence, which the latter had constructed to separate his land from an Esplanade which had been built in 1859.

The plaintiff alleged that his access to the highway was obstructed due to construction of this fence but the defendants said the land was not a public highway. However, the people of the village had been using the land as a highway for years to get to the promenade.

Thus the defendants were compelled to remove the fence and pay damages - to be determined by a separate enquiry.

11   **A-G. v Blackpool Corporation [1907] 71 JP 478**

The local authority gave permission for motor car racing on the promenade ('The Parade') to be properly policed and controlled. The legislation which gave power to the authority to build the promenade restricted use to 'foot passengers, perambulators, invalid carriages, and similar vehicles'. Was the authority acting within its powers (intra vires) when it gave permission to hold the race?

Held: The authority were in the position of trustees of the parade for limited public purposes... and that it was an abuse of 'The Parade' to allow it to be used for either horses or motor cars and a fortiori motor races. It was therefore acting ultra vires, ie outside its powers.

12   **Roberts v Webster (1968) LGR 298-305**

Appeal case stated by Cheshire Quarter Sessions who on 9 January 1957 dismissed an appeal by the present appellant William Roberts

of Wirall Urban District Council, from a decision of the Wirral
Justices that Pipers Lane at Heswall in Cheshire was a highway
repairable at public expense.
A street works authority served notices on the frontages of a lane
requiring them to undertake repairs under section 204 of the 1954
Highways Act. The frontages objected that the lane was repairable
at public expense and section 204 did not apply. The authority
conceded that at the present time this lane was a public highway
having been used as a right of way by the public for a substantial
period of time, but contended it had achieved status after 1835. The
lane about three-quarters of a mile long started from an ancient
highway and ran parallel to a shore estuary, terminating in glebe
land near a natural beauty spot.

Maps showed it had existed as a road since before 1835, and it was
shown, inter alia, on an enclosure award made in 1859 under the
1845 Enclosure Act. No public money had ever been spent on it.
Quarter Sessions, concluding that the enclosure award of 1859 was
such powerful evidence that they should infer from it that a
highway existed over the lane in 1859, and that it was not shown
that it had not existed before 1835, dismissed an appeal from a
decision of the justices upholding the objections.

13   **Moser v Ambleside UDC (1925) JP 118-120**

In 1834 the tenant in tail of the land over which an alleged highway
ran barred the entail and betwen then and 1842 was capable of
dedicating. In 1842 he conveyed the land to trustees upon a
settlement which was assumed to be strict. In 1875 he died,
bequeathing the land to trustees for sale. In 1879 the trustees sold
the land and the purchaser immediately mortgaged it to the
trustees of the will of another testator. In 1899 the plaintiffs, the
trustees of a third testator, bought the land, and from then until the
date of the action the land was admittedly in strict settlement. Two
small maps each 100 years old showed a road on the line of the
alleged highway. Held upon the evidence, the plaintiffs had failed
to establish that there was no right of way.

14   **The Board of Works for the Greenwich District v Maudsley and
      others (1870) LR 5QB 397**

A seawall or river wall passed through the defendants' land. It had
existed as long as the records went back. The local people relied
upon the path along the top of the wall as a means of
communication between factories and buildings and the adjacent
neighbourhood, and from one factory or building to another. It was

also used as a pleasure walk and has also been used without interruption by watermen and pilots. It was at several points along the course of the path divided by fences. Stiles were put up and maintained. In August 1865 the defendants stopped up the path by erecting across it a high wooden fence backed by a quantity of iron. This was removed by the plaintiffs but was immediately replaced by the defendants.

Is the defendant guilty of obstructing a public way? Yes!

If a right of way has been used for a great number of years it is deemed to be a public right of way.

Before constructing on any footpath it should be checked as to whether or not it is a public right of way. If so, do not obstruct it.

15   **Farquhar v Newbury UDC (1909) 1 Ch 12 CA**

The action was brought for a declaration that a roadway across the Shaw House estate, near Newbury, was not a Public Highway. The plaintiff was the owner and occupier of the Shaw House estate, which was situated between the villages of Shaw and Donnington. Since the year 1632 these villages formed one parish with one church, St Mary's, which was close to Shaw House and the village of Shaw. The plaintiff, who bought Shaw House in 1905, admitted that the inhabitants of Shaw and Donnington were entitled to use the road as a churchway for church purposes, but denied that the public had any right to use it, and in order to assert her right she locked two gates which stood across the road. The defendant council disputed her right to close the road and removed the locks, whereupon the plaintiff took action.

The court held: Specific intention to dedicate must be shown, and it is that which differentiates this case from the long line of authorities where dedication has been presumed from long user, there is here no evidence that Dr Penrose, a previous owner, never knew that this road was going to be dedicated to the public and in fact dedication was not in Eyres mind (his nephew and estate manager).

16   **A-G v Mallock (1931) ALL ER 204**

Mr Mallock, the vicar, interrupted a way over his land to the historic Cockington village church by putting up notices to deter people who were not visitng the church. The A-G. claimed a declaration that the way was a public highway and sought an injunction to prevent 'M' from interfering with the public use of the alleged right-of-way. He argued the highway was dedicated by a previous owner at a time unknown. He claimed a presumed dedication from uninterrupted use by the public throughout living memory and reputation.

For a highway to exist there must be a dedication either express or presumed. There is, in this case, no express dedication and the erection of notices over a long period of time, restricting access to the times when the church was open is an indication of the lack of it. A quotation from Poole v Huskinson (1843) shows its importance '... a single act of interruption by the owner is of much more weight, upon... intention, than many acts of enjoyment'. Access to the church was controlled by the vicar; the church is vested in him ('vicars freehold').

The action was dismissed.

17   **Cubitt v Lady Caroline Maxse (1873) LR 8, CP704, Hals Vol 21**

Effingham Common was enclosed by statute in 1802. In 1808 a highway was authorised but never built, trees were planted on the line of the highway and when the owner cut them down in 1869 he was sued in trespass by an adjacent owner; the action was justified because the trees were on the highway. Held: statutory authorisation insufficient, it must be set out or substantially completed.

18   **Austins Case (1672) I Vent 189**

'If it be a public way, of common right, the parish is to repair it unless a particular person be obligated by prescription or custom.'

19   **Russell v Men of Devon (1788) 2 Term Rep.667, 26 Digest (Repl) 654, per Lord Kenyon C.J.**

Old rule of law on maintenance of highways. The special defence of non-feasance available to highway authorities or historically the surveyor when sued for damages following an accident arising out of a failure to maintain the highway.

This defence was abrogated by S I Highways (Miscellaneous Provisions) Act 1961. S44 of the 1959 Act and now S41 of the 1980 Act place this obligation to maintain on the local highway authority.

Nonfeasance: The neglect or failure to do some act which ought to be done e.g. failing to repair the highway (distinguished from misfeasance, i.e. improper performance of a lawful act,such as negligence or trespass. eg negligently repairing the highway).

20 Thompson v Mayor and Corporation of Brighton (1894) 1QBD332

The plaintiff was riding along a public road in Brighton. His horse's foot struck the cover of a manhole in the middle of the road, which projected about one and a half inches above the surface of the road, and the horse was thrown down and seriously injured. The manhole had been inserted in the road by the corporation of Brighton as the sewer authority. It was a proper cover, and there was no fault in its construction nor was it at all out of repair. The accident arose from the road not having been kept up to its level by the corporation, who were the road authority.

The plaintiff sued the corporation of Brighton in the county court for £50 as damages occasioned by their negligence.

The county court judge, in opposition to his own opinion, gave judgement for the plaintiff for £50 considering himself bound by 'Kent v Worthing Local Board' in which his own decision in that case had been reversed. The defendants are sued, not as the sewer authority, in which capacity they were guilty of no default, but as the road authority.

21 Skilton v Epsom and Ewell UDC (1937) 1 KB 112

'S' whilst cycling arrived at a line of studs in the road at the same time as a car. As the car overtook her it ran over a loose stud which was propelled at such speed that it knocked 'S' off her cycle as a result of which she was injured.

The studs had been fitted by the council under the authority of the Road Traffic Act 1930; there was evidence the stud had been faulty for some time. The action was for personal injury. The council claimed the stud was part of the highway and that misfeasance for which the highway authority could be liable had not been proved.

Held: Assuming that the stud was part of the highway and that it became defective and amounted to a nuisance on the highway; the LA could not take advantage of cases dealing with maintenance of highways because the stud was not brought on to the highway as part of the maintenance of it under the 1835 Act, but for the purpose of traffic direction under the Road Traffic Act 1930.

22 Macclesfield Corporation v Great Central Railway (1911) 2 KB 528

'MC' repaired a highway which was carried by a bridge in the ownership of 'GCR' over a canal. The road was in disrepair and had become dangerous, 'MC' only doing the repair following their request to repair to 'GCR' who refused. The action is to claim the cost of the repairs.

Held: (i) 'GCR' were responsible for repair to highway and the fabric of the bridge; (ii) the plaintiffs had no legal liability to repair and had acted as volunteers and could not recover the cost of repairs from 'GCR'.

23 R v Baker (1980) CL Review April 22, p.600

The appellant pleaded guilty to assault occasioning actual bodily harm.

An argument broke out between the appellant and another man as a result of the appellant's car blocking a driveway (access to a dwelling), and the appellant punched the other man, fracturing his chin and jaw.

The appellant claimed that the victim had started the violence.

The appellant had previously had three convictions and fines for violence and in this case was found guilty. Sentence 12 months imprisonment, suspended for 18 months and a fine of £200, payable at £10 per week, with three months imprisonment for default.

24 Rundle v Hearle (1898) 2 QBD 83

Appeal from the county court of Cornwall. The defendant was the occupier of the farm through which ran a public footpath, which crossed a fence separating two of the defendant's fields by means of a stone stile. Owing to the stones of the stile being worn away the plaintiff, who was using the footpath as a member of the public, fell in attempting to get over the stile and was injured. The plaintiff brought an action for damages against the defendant, alleging that he was liable to repair the path and stile ratione tenurae. The defendent had been in occupation of the farm for thirty years. About twenty years before the plaintiff's accident a servant of the defendant had repaired the path and put a new stone step in the stile. The defendant's predecessor in the occupation of the farm was a Mr Martin, who was also the owner, and there was evidence that he used to repair the path on both sides of the stile. Martin's predecessor was a man named Hockling, who had occupied the farm for thirty or forty years. He used to repair both the pathway and the stile. It was objected before the county court judge that these facts afforded no evidence of any liability in the defendant to repair ratione tenurae (as an obligation of the tenancy), and it was also contended that even if the defendant was so liable, no action would be against him for a non-feasance. The county court judge overruled both objections, and gave judgement for the plaintiff. The defendant appealed.

First, the facts proved afford no evidence of any liability on the defendant to repair ratione tenurae; the defendant is the occupier and not the owner and is liable. Secondly, if the defendant is liable to repair, no action will be against him for the omission, the only remedy is indictment.

As to the proposition that no action will lie for non-repair, the cases quoted are not in point, for they show that a public body can't be sued. The reasons why such an action will not be against a public body have no application to the case of an individual. Appeal allowed.

25   **The Mayor of Tunbridge Wells v Baird and Others (1896) AC 434**

The council built public conveniences beneath the street under the authority granted by a local Act. 'B' and others were owners of the freehold on either side of the 'Pantiles' (a street where people promenaded and partook of the waters) and sought a declaration that the ground under the highway was in their ownership and a mandatory injunction to remove the works.

Held: The authority had no power to excavate the soil and erect lavatories below the surface of the street which had vested in them within the meaning of the Public Health Act 1875.

26   **Tottenham UDC v Rowley (1912) 2 Ch 633; CA**

The defendant laid out and developed his property as a building estate in accordance with a plan approved by the plaintiffs. The defendant built houses on the southern side of the road overlooking the plaintiffs' boundary, and made up and metalled the road for one half of its width next to the houses. The further or northern half of the road next to the plaintiffs fence was left unmetalled and untouched. The defendant subsequently removed the fence from the end of the road, and the evidence showed that for some three years before action was taken, the road was used as a thoroughfare by pedestrians, cyclists and vehicular traffic, unconnected with residents in the road.

The plaintiffs had recently opened a gate in their boundary fence and were carting building materials across the unmetalled portion of the road. The defendant objected on the ground that the unmetalled portion was not subject to any public right of way, and that even if it were, the plaintiffs were not entitled to take their carts over the part appropriated solely for the purposes of a footpath.

27   Marriot v East Grinstead Gas and Water Company (1909) 1 Ch 70

The plaintiff brought about the action that (1) the defendants were not entitled in connection with the said works of Hockenden to lay down or place any pipes, conduits, service pipes or other works in the plaintiffs' footpath. (2) Alternatively a declaration that the defendants were not entitled to lay such pipes or other works without giving notice to and obtaining the consent of the plaintiffs. (3) An injunction.

It was held that the defendants were not entitled to lay a 6' pipe or other works in the plaintiffs' footpath.

Also notice must be given and the plaintiffs' consent obtained for work to be carried out on the west side of the plaintiffs' property.

An injunction restraining the defendants, their servants and agents from laying down or placing a 6' main or any other pipes or works in connection with their works and pumping station at Hackenden under the said footpath.

28   Wood v Ealing Tenants Ltd (1907) 2 KB 390

'W' owned a house and garden adjacent to a passage owned by 'ET'. Beneath the passage (in private land) there was a pipe which as a result of it draining more than one building was defined under the Public Health Act 1875 as a public sewer. 'W' had his drainage connected to this public sewer by the LA. 'ET' later removed the pipe (one foot long) from the boundary wall to the sewer. 'W' sued for trespass.

Held: Whilst 'W' was entitled to connect to the sewer neither he nor the LA had the right to place 'W's pipe in 'ET's' land.

29   Porter v Ipswich Corporation (1922) 2 KB 145

'Ip.C' erected four poles, to carry electric wires, on a banking which 'P' claims is his property and 'Ip.C' claim is part of the highway, 'Corp' have also erected two poles in the highway digging to a depth of 6 ft. In neither case did the council seek 'P's' permission as landowner. 'Ip.C' claimed the bank is part of the highway and that therefore they have the authority to do the work.

Held: The bank as a matter of fact was 'P's' land, so trespass had occurred. On excavating 6 ft deep legislation allowed the utilities gas, water, sewers, electricity to 'break open' the highway to fulfil their functions. If permission from landowners was required it could make legislation inoperative.

30   Fenna v Clare and Co (1895) 1 QB 199

The defendants were the owners of a shop adjoining a public highway. In front of a receding window of the shop, and immediately abutting on the footway, was a low wall 18 inches high, the property of the defendants. On top of which wall was a row of sharp iron spikes 4½ inches high. The plaintiff, a little girl aged 5 years 9 months, was found standing with her arm bleeding from a recent wound, such as might have been caused by her falling upon the spikes. No-one witnessed the accident. The plaintiff brought in action in the county court for damages for the injury, which was alleged to have been caused by reason of the defendants having maintained a state of things which was a nuisance to the highway. At the trial the plaintiff who was ill was not called as a witness, nor was any other evidence other than the above given as to how the accident happened, except that of the defendants' shop boy, who saw the plaintiff climbing the wall before the accident and told her to get down which she did. The defendants' counsel submitted, upon the authority of Wakelin v London and South Western Railway Co, that there was no case to go to the jury: for that, assuming that the spiked wall was a nuisance, and that the plaintiff's injury was caused by her falling upon the spikes, the facts proved were as much consistent with her falling whilst wrongfully climbing upon the wall in which case she should, upon the authority of Hughes v Macfie (2), be disentitled to recover as with her having fallen whilst lawfully using the highway. The county court judge however refused to non suit, and left to the jury the following questions:

1 Were the spikes a nuisance - answer YES
2 Was the injury caused by the spikes - answer YES

3 Was there contributory negligence on the part of the plaintiff - answer NO

Upon these findings the judge entered judgement for the plaintiff. The defendants appealed. Appeal dismissed.

31   **Harold and Another v Watney (1898) 2 QB 320**

The plaintiff (4 years of age) was attracted by some older boys playing on the other side of the fence. He put his foot on it in order to climb over when the fence fell on and injured him. In the action for damages for the injuries sustained, the jury found that the fence was very defective, but actually fell through the plaintiff standing partly on it, though not for the purpose of climbing over.

Held: That the defective fence was a nuisance and the cause of the injuries to the plaintiff and so the defendant was liable.

There are a number of cases to show that the trespasser may have the right of action for an injury sustained while in the act of trespassing.

32   **Bromley v Mercer (1922) 2 KB 126 TLR CA**

A young girl whilst staying with her grandmother was injured whilst playing, after putting her hand on a stone in a 3' 6' wall whose iron holding stays had rusted away, the stone fell on her, causing sufficient injury to warrant her leg being amputated.

An action was brought against the landlords who had covenanted to keep the premises in structural repair and it was found that they were liable due to the condition of the wall constituting a public nuisance to persons using the highway on the opposite side of the wall.

The defendants appealed and won with the ruling that, 'there is nothing to show that an invitee upon private premises is entitled to damages for an injury caused by something which is a public nuisance to an adjoining public highway.' The landlords could not be sued as they were not in occupation of the premises.

Any member of the public has the right to recover damages for injury caused by a public nuisance and from those who permit the public nuisance to continue, as long as they are lawfully exercising a public right to use the highway. This right is not extended to private premises.

33   **Caminer and Another v Northern and London Investment Trust Ltd (1951) AC 58**

On 7 April l947, appellants were driving past some flats when a diseased tree fell on to their car wrecking it and injuring them in the process. The tree was old, diseased and uncared for. The appellants tried to prove the owner negligent and a nuisance to the public: was the landowner liable and did they have ground to claim compensation.

The House of Lords stated that although it was good practice to operate a decent 'estate management' policy, and that the respondents (landowner) obviously didn't or hadn't; they the respondents were laymen and as a result were ruled not to be liable.

34   **Leasne v Lord Egerton, (1943) 1 ALL ER 489**

On the night of Friday 15 November l943 there was an air raid, which caused considerable damage to the window frames and glass of an empty house owned by Lord Egerton. Lord Egerton had personal knowledge of what happened but no steps were taken to resolve the damaged property until the following Monday. This was due to the fact that the offices of agents were closed on the Saturday and there was difficulty in getting builders on that day. Before the repairs took place the plaintiff, Mrs.Leane, was walking past the premises, when a piece of glass fell and hit her sustaining serious injury.

Principle of law: The house in its damaged condition was a 'nuisance' and the defendant was liable for continuing the nuisance of which he had 'presumed knowledge'.

Decision: The owner continued the nuisance therefore was prosecuted for negligence and hence the plaintiff was awarded £275 with costs.

35   **Crane v Southern Suburban Gas Company (1916) 1 KB 33**

Workmen employed by the defendants, for the purpose of carrying out repairs to a gas main in a highway, placed a fire pan, on which was a ladle containing molten lead, on enclosed land adjacent to a highway. The plaintiff, a young child, was playing with other children near the fire when a passer-by accidentally knocked over the fire pail, and the molten lead was spilled on the plaintiff, causing her injury. In an action by the plaintiff to recover damages the county court found the defendants were guilty of negligence in having the fire unattended and unguarded with the knowledge that it was surrounded by children and that it was being used for molten lead.

36   Halsey v Esso Petroleum Co Ltd (1961) 2 ALL ER 145

The plaintiff occupied a house in a road zoned for residential purposes. Esso had an oil distributing depot adjoining the plaintiff's road in an area zoned for industrial purposes. Oil was pumped from river tankers on to the depot and then into road tankers. The medium and heavy grades of oil needed to be heated in order to be pumped. The throughput of oil had almost doubled in a four year period up to 1957. Therefore in 1956 night shift working was re-introduced. On the depot and opposite the plaintiff's house was a boiler house with two metal chimneys used to heat the fuel oil. From these chimneys acid smuts containing sulphate were emitted and were visible falling outside the plaintiff's house. There was proof that the smuts damaged clothes hung out to dry in the plaintiff's garden and the paintwork of the plaitiff's car parked on the highway outside his house. The depot emitted a smell which would affect a sensitive person. The plaintiff suffered no problems of health from the smell. During the night the noise from the boiler caused windows and doors to vibrate. The defendants had soundproofed the walls of the boiler house but to no noticeable effect. Also the large road tankers entering and leaving the depot during the night caused considerable noise while they manoeuvred on the public highway on entering the depot. All of these factors of noise prevented the plaintiff from sleeping.

Held: The defendants were liable to the plaintiff in the following respects and on the following grounds:
(i)   for the emission of acid smuts
    (a) in respect of damage to clothing and car since it was caused by harmful substances escaping from the defendants' premises;
    (b) as a private nuisance, damage was a material injury to the plaintiff's property resulting by a trade carried on by the defendant;
    (c) as a public nuisance, damage was caused to his car while on the public highway;
(ii)   in respect of nuisance by smell because the smell emanating from the defendants' premises amounted to a private nuisance, injury to health was not a necessary ingredient in the course of action for nuisance by smell;

(iii) in respect of a private nuisance by noise from the boilers and
       road tankers when in the depot;
(iv)   for the noise from road tankers made at night on the public
       highway
       (a) as a public nuisance, since the concentration of moving
           vehicles in a small area of the public highway was an
           unreasonable user of the highway;
       (b) as a private nuisance, since the noise was directly related
           to the operation of the depot.

37   **Lagan Navigation Company v Lambig Bleaching (1927) AC 226**

By a local Act of 1843 a canal, constructed under previous local
Acts by canalising a river was vested in the appellants. They were
required by the Act to keep the navigation, locks and any improvements in an efficient state for the traffic using the canal.

In 1912 the appellants raised the copings on both sides of one of
their locks and the banks behind it to prevent the lock from being
flooded. The respondents, adjacent landowners, objected that the
effect of these works was to pen back the water in the part of the
canal above the lock and to occasion the flooding of their lands. In
consequence of these objections, the appellants removed the
copings but not the banks, without prejudice to their rights. In 1924
during a heavy flood, the respondents, without notice to the
appellants, cut away a portion of these banks to allow the flood
water to escape.

In an action by the appellants for an injunction to restrain the
respondents they justified their acts as having been done in the
abatement of a nuisance caused by the appellants in raising the
banks. Held that (1) there being no evidence of negligence, the
appellants, in constructing works in the exercise of their statutory
powers for the protection of their navigation,were not liable for the
flooding of the respondents' lands; (2) that, assuming that the
raising of the banks and the resultant flooding constituted a
nuisance, the course pursued by the respondents was not justified.
Lord Wrenbury deciding the case on a view of the facts presented.

38   **Dovaston v Payne Blackstone's Reports Common Pleas
       1788-1796 Occurred on 10 January 1795**

The facts are that the cattle of the defendant escaped from the
highway to the land adjoining. They escaped through an
unrepaired fence which the owner should have repaired but failed

to do so. The cattle may be distrained if they were not lawfully upon the highway. The cattle then grazed on the plaintiff's property.

Eyre, C. J. said he agreed with the counsel for the defendant as to general law that the party who would take advantage of fences being out of repair as an excuse for his cattle escaping home using a way over the land of another must show that he was lawfully using the easement when the cattle so escaped.

This case is different from cattle escaping from a close, where it is necessary to show that the owner had a right to put them there because the highway being for use of the public, cattle may be on the highway of common right.

39 Hickman v Maisey (1900) 1 QB 752

'H' was the owner and occupier of land on the Wiltshire downs which was crossed by a highway, the highway being used for training racehorses. 'M' was the proprietor of a racing journal commenting on and giving tips on the horses. 'M' would often observe the horses from the highway for this purpose. The trainer objected and gave notice to 'M' for him to stop using the highway for this purpose; 'M' refused. Action was taken to sue for damages for trespass and an injunction to prevent 'M' repeating his conduct. The judge directed the jury ' "M" as one of the public was entitled to use the road for the purposes of passing as a traveller; but not to stop for the purpose of his own business...' On the evidence the jury decided it was an unreasonable use of the highway. Damages of one shilling (5p) were awarded and an injunction granted.

40 Harrison v Duke of Rutland (1893) 1 QB 142

The defendant was the owner of a moor across which ran a highway, the ownership of the soil of which was in the defendant. The defendant on an occasion was with a party of friends, lawfully engaged in shooting grouse on land adjoining the highway, the plaintiff, for the express purpose of annoying the defendant went on to the highway and tried in various ways to divert the grouse, which were being driven towards the highway and in the direction of the butts where the defendant and his friends were stationed. The plaintiff refused to cease what he was doing, and certain keepers of the defendant, by his orders, held the plaintiff down on the ground till the drive was over. In an action by the plaintiff for assault and false imprisonment, the plaintiff was a trespasser on the highway, and a declaration would be made to that effect.

When a highway is dedicated to the public the easement acquired by the public is a right of passage, a right to pass and repass along the highway at their pleasure for the purpose of legitimate travel; conduct other than that, whether lawful or unlawful, is an infringement of the rights of the owner of the soil who has not transferred the absolute property of the soil of the highway but, subject to the easement, has precisely the same estate in the soil as he had before the easement was acquired by the public and constitutes the person using the highway for a purpose other than exercise of the right of passage a trespasser as against the owner of the soil on the highway.

41 Rogers v Ministry of Transport (1952) 1 ALL ER 634

Lorry drivers parked on grass verge whilst having a break in a cafe. The grass verge is part of highway and should they drive over it they would not be doing anything illegal. The authority constructed a layby for lorries to park; Rogers owned adjoining property and wished to prevent the layby being constructed. If it is part of highway it may be used for legitimate purposes, eg a temporary stop and must not cause an obstruction (this is within normal uses of the highway). There may also be local bye-laws applicable.

42 R v Surrey CC Ex p Send Parish Council (1979) 40 P & CR 390

A path, which on the evidence had for many years been used as a public footpath, became increasingly obstructed by fences, buildings and the like erected by owners of the abutting land ('the landowner'). From 1969 the parish council and other local organisations and residents sought to persuade the respondent highway authority to take action under S116 of the Act to prevent the continued obstruction. Protracted negotiations took place culminating in the landowner's in 1977 issuing a writ seeking a declaration that the path was not a public footpath. No further action was taken. The authority having previously rejected the landowner's proposal for resiting of the footpath eventually agreed there to despite local opposition. The parish council claimed that the authority was in breach of its duty under S116 (6) 'to take proper proceedings' for the removal of the obstructions.

A highway authority may be compelled by mandamus to take action to prevent the continued obstruction of a public footpath.

Held: Granting an order of mandamus, that the authority's inactivity could not be excused in the light of their clear duty under Sll6 and that although they clearly had a discretion as to the nature of 'Proceedings' to be brought, some action upon their part was required.

The case illustrates the awareness of footpaths required when setting out a site. Investigate whether a footpath is privately owned or for public use, and if it can be diverted or closed. Footpaths may sometimes be diverted or resited but it may have the right of way.

Setting out may affect choice exits and/or entrance to a site; fences around a site might obstruct a footpath.

43   Noble v Harrison (1926) 1 KB 332

The branch of a tree growing on 'H's land and overhanging the highway (30 ft) fell off on to the passing van of 'N' causing damage to the vehicle. In the county court it was determined that it was a latent defect in the tree which caused the accident and which was not discoverable by any reasonable inspection; but he found 'H' liable on the rule from Rylands v Fletcher (l868) and nuisance (overhanging branch). 'H's appeal was upheld: a tree was a natural thing to have on land and therefore 'R v F' did not apply and that the branch overhanging the highway was not a nuisance without it being an obstruction. The branch was a danger but 'H' was not liable as he had not created nor was he aware of and had no knowledge actual or imputed of the existence of the danger.

44   Marshall v Blackpool Corporation (1935) AC 16

'M' applied to the Corporation under a local Act for permission to build a carriage crossing from his motor coach garage to the road outside. All the construction details were correct, but permission was refused on the basis of safety of the general public in this location. 'M' appealed.

Held: Under the statute, the authority did not have the power to refuse permission on these grounds. At common law, the owner of land adjacent to the highway has a right of way to the highway.

45   Cobb v Saxby (1914) 3 KB 822

'C' and 'S' were the respective owners of adjoining houses and shops in Margate, adjacent to the highway. The side wall of 'S's house projected into the street beyond 'C's house. There was no door or window in the wall. 'C' fixed boards and advertisements to this wall (l6" x 22 ft) preventing 'S' (the owner) access to it from the street for either repairs or advertising of his own.

Held: The right of access includes a right of access to the wall in which there is no door. The action was an infringement of 'S's right of access and he was entitled to an injunction. (The action originally brought by 'C' because 'S' had placed his advertisements on the boards which 'S' acknowledged were owned by 'C', this remaining action is 'S's counterclaim.)

46   **Perry v Stanborough (Developments)(1977) 244 EG 551 Digest CLY2502**

'P' bought land without rights of access to the highway assuming that he would obtain access over the intervening land by virtue of the fact that the owners had obtained planning permission to develop it on condition that an estate road be built up to the boundary. The road was constructed and surfaced up to a point short of the boundary and was adopted by the local highway authority. 'P' argued they should either insist on the road being built up to the boundary or obtain the land by compulsory purchase.

Held: There was no duty or planning authority to force or secure extension of road or to aquire the land by compulsory purchase; the matter was one for private negotiation between the adjoining landowners.

47   **Barber v Penley (1893) 2 Ch 447**

The lessee of a theatre was held liable for obstruction because of the crowds gathering on the footpath prior to the earlier performances of *'Charley's Aunt'*. The crowds prevented access to adjacent property, a common lodging house.

Held: There was an obstruction at the time the action was brought but it had since ceased following action by the police. 'B' was granted costs.

48   **Leonidis v Thames Water Authority (1979) 77 LGR 722**

The plaintiff had a motor repair business which led off a major road (Melbourne Square). It was a two-way street and thus persons wishing to resort to it could, in normal circumstances conveniently do so.

Between May 1975 and April 1976 the defendants, exercising their statutory powers, closed off the access from the major road in order to reconstruct a sewer at the junction with the side street. Although there was an alternative route to the garage, that too, was closed off by the defendants between July and September 1975. As a result, access from the plaintiff's premises to and from Brixton Road was, during the period of 11 months, entirely impossible for vehicular traffic. During this period, any other route to the garage required a long detour.

As a result of the defendants' operations, the plaintiff considered that the profits from his business had suffered and he made a claim against the defendants for compensation under section 278 (1)of the Public Health Act 1936, which provides:

'Subject to the provisions of this section, a local authority shall make full compensation to any person who has sustained damage by reason of the exercise by the authority of any of their powers under this Act in relation to a matter as to which he has not himself been in default.'

The defendants disputed liability. The matter was referred to a single arbitrator under the procedure prescribed by the Act.

Held: Giving judgement for the plaintiff for the amount claimed;
(1) that loss of profits was recoverable (£2,500)
(2) that the fact that the entrance to the plaintiff's premises was not obstructed was not relevant
(3) whilst the arbitrator's award could not be conclusive in law and an award might be bad for error on its face, nonetheless, there was nothing to suggest that the arbitrator had no power to determine liability.

Accordingly, there must be judgement for the plaintiff for the amount claimed and, by agreement, an order for the plaintiff's costs.

49   **Lyons, Sons & Co v Gulliver (1914) 1 Ch 631, CA**

The owners of the Palladium Theatre, the defendants, gave performances of various plays which were shown in the mornings, afternoons and evenings. The people going to the theatre caused a queue which extended for a considerable distance in front of the neighbouring shops. The defendant's however kept the front of the theatre door clear thus allowing the queue to extend further down the street. Sometimes the queue being on the footpath and other times on the roadway. The plaintiffs, Messrs Lyons Sons & Co who owned a shop a few yards from the theatre, complained that the stationary queue prevented their customers from reaching the shop. They said that this resulted in lost custom as their customers found it difficult to reach the shop because of the queues. The plaintiffs stated that the theatre doors were not open until about a quarter of an hour before the performance thus extending the time the queue was outside the shops, however it was said by the defendants that it was the duty of the police to keep the streets clear.

The point of law to be considered is: that there was an obstruction of access to the neighbouring shops by the theatre crowds, and that the occupier of the theatre premises may be liable for a nuisance even by something lawfully done if his neighbour suffers damage. The nuisance depends on the degree of annoyance and obstruction caused.
Held: The defendants were liable to the plaintiffs for the nuisance caused.

50   **Harper v Haden (1933) 1 Ch 298**

The defendants were constructing a new storey to a building in which they were the occupiers of the upper floors. The plaintiff owned a shop on the ground floor. Before construction started, scaffolding and hoarding were put up which rested on the footpath. Before the scaffolding and hoardings were placed, the necessary licence was obtained for their erection from the surveyor of the Holborn Borough Council. On 28 August 1931, plaintiff commenced action for injunction to restrain defendants obstructing the pavement by scaffolding and hoarding. This was refused. The plaintiff proceeded for trial on the question of damages.
It was a matter of give and take. To be obstructed occasionally was the price one paid for the privilege of obstructing others. Obstruction to highway may be legal if obstruction is temporary in character and reasonable in quantum and duration.
The highway may be used in the construction industry eg for the erection of scaffolding if within reason and temporary in character.

51   **Milward v Redtich Board of Health (1873) 21 WR 429 Digest (Repl) 351**

A local board of health in repairing and improving a road under the powers of Public Health Act 1848 raised a footpath by the side of the road a few inches, the effect of which was to prevent water which fell upon the space between a warehouse of the plaintiff in which needles were stored and the road from draining into the road.

On a bill filed by the plaintiff against the local board for an injunction:

Held:

(i)   they had no rights to make improvements in a way calculated to cause unnecessary injury to the plaintiff;

(ii) the evil complained of was one of easy remedy;

(iii) the case was not one of pecuniary compensation and a mandatory injunction would be granted to prevent defendants from allowing such water to remain dammed up to the injury of the plaintiff.

52   Hawkins v Minister of Housing and LG (1962) 14 P & CR 44

Planning permission was granted subject to a condition limiting vehicular access to single access between specified points, and requiring the design and construction of the access to be approved by the planning authority. The property comprised some 3½ acres of land to the rear and on both sides of the claimant's residence in Boroughbridge.

A claim for development value under the Town and Country Planning Act of 1947 was lodged and an unrestricted value of £1,500 and a restricted value of £600 was determined. The restrictions imposed by the Minister, as Colonel Hawkins alleges, caused losses of £908. 8s 7d.

The Ministry of Housing and Local Government opposed the claim on two grounds. First there was no depreciation of the property by virtue of the conditions made by the Minister, and he called as witness Mr Henderson of the R & CS. In his opinion there was no great demand for houses in Boroughbridge and the new by-pass would deteriorate trade and, consequently, the development of the claimant's property unless a proper road access as required by the Minister to the development was possible.

53   Wolverton UDC v Willis (standing as S G Willis & Sons) (1962) 2 ALL ER 243

Mr Stanley Gordon Willis was charged on 1 February 1961 for displaying vegetables which projected onto or over a footway a distance of 11 inches beyond the line of his shop, No. 45 High Street, Stratford, where they were said to obstruct or incommode the passage of any person walking along this footway. This was contrary to the Town Police Clauses Act 1847 (Section 28).

The judges were of an initial opinion that the display of vegetables did not constitute anything more than a mere technical obstruction and it would not obstruct the passage along the footway and dismissed the summons on the 'de minimus' factor. However, an appeal was lodged and the greengrocer was guilty of the offence charged since the 'de minimus' principle was wholly inapplicable to such encroachment, and there was no need to allege that passage had been obstructed.

54   Seekings v Clarke (1961) 59 LGR 269

In the summer of 1960, a shopkeeper who conducted a tobacconists, newsagents and fancy goods business in a seaside town in a road leading down to the sea erected a sunblind in front of his shop. The sides of the blind, to which were attached wire displays containing books, projected 2 ft 6in over the pavement which was 16 ft wide. He also placed wooden shelves containing books, 9ft by 6 ft 6 in, on the pavement in front of the shop and articles for sale, such as buckets and spades, were set out on the pavement between the bookshelves and the window displays. The shopkeeper was guilty of wilfully obstructing the free passage along the highway contrary to Section 121 of the Highways Act 1959.

55   Lowdens v Keaveney (1903) 2 R 82

The court ruled that there may be a considerable, even complete obstruction of the highway and yet the use might be reasonable; for example in the case of a procession.

56   Castle v St Augustine's Links (1922) 38 TLR 615

According to the evidence, the plaintiff George Castle on 18 August 1919 was driving a taxi from Deal to Ramsgate when a golf ball played by Mr Chapman from the 13th tee on the golf course parallel with Sandwich Road struck the windscreen of the cab. A piece of glass from the windscreen injured Mr Castle's eye which had to be removed. At the time he did not know of the existence of the golf course.

William S Castle, the brother of the plaintiff said that he had spoken to Mr Chapman on the day of the accident and Mr Chapman had admitted that it was him who drove the ball. The witness said he drove past the golf course 20-30 times a day but had no reason to complain, because he had never been hit. But he did reckon that the 13th tee was a danger because he had seen boys retrieving golf balls from the road. The question is how much if at all was the golf club liable? It was alleged that the golf club was guilty of a nuisance by setting out the course in this way and not providing any warning signs. His lordship gave judgement for the plaintiff of £450 against both Mr Chapman who had gone to Australia and the golf club, but said that he could not make the golf club pay thousands of pounds for such a regrettable accident.

Principles of law involved should the golf club be held liable for an accident that they did not really take part in or only took part in in a secondary way.

Held: Plaintiff was awarded £450 damages.

57   Bolton v Stone (1951) AC 650, ALL ER 1078

Miss Stone was standing outside her house, 10 Beckenham Road, on 9 August 1947 when she was hit by a cricket ball and injured. The ball was hit from the cricket ground adjacent to the road, by a batsman playing in a match. The ground is owned by Cheetham Cricket Club, the respondent brought an action for damages against the committee and members of the club, not the individual striker of the ball.

The distance from the striker to the respondent was about 100 yards. The ball was hit over a protective fence rising 17 feet above the cricket pitch and 78 yards from the batsman. The hitting of a ball onto the road has only been proven to have occurred 6 times in over 30 years. No-one had been known to be injured.

Miss Stone obtained damages for negligence and nuisance by the members of the committee of the Cheetham Cricket Club. She claimed that sufficient care was not taken by the club to ensure that a ball would not be hit out of the ground and injure anybody on the adjacent road.

The defendant would be negligent if it was reasonable to think that the ball may be hit out of the ground and hit a passer-by, and they knew this and did not take any precautions to stop this happening. In fact, it would be desirable as part of the game of cricket to try to hit the ball out of the ground.

As the hitting of a ball out of the ground is a very rare occurrence, as it was known to be, it is not reasonable to think that any reasonable man would know the ball would be hit out of the ground and injure someone, to warrant him to be negligent. Nuisance cannot be established if negligence is not proven.

Held: The risk of injury to a person on the highway resulting from the hitting of a ball out of the ground was so small that the probability of such an injury would not be anticipated by a reasonable man, and, therefore, the members of the club are not liable to the respondent.

58 Miller and Another v Jackson and Others (1977) 3WLR 20; QB 966

Members of a Cricket Club (Jackson and others) had played cricket on a ground rented from the N.C.B. since 1875. In 1965 the N.C.B. sold the adjacent land to a private developer. In 1970, the ground opposite the cricket field was built up and in 1972 the plaintiffs (Millers) bought a house on the periphery of the cricket ground. As time went by, cricket balls were hit onto the Miller household's garden, and caused damage to their property. They complained, subsequently, the cricket club erected a large fence. Damage still occurred and the Millers complained that it might result in someone being personally injured.

The cricket club offered to place shutters, unbreakable glass etc, and pay for any damage against them. The Millers refused this and decided to go to court to obtain an injunction, to be placed against the cricket club, to prevent them from playing cricket, on the grounds of nuisance, damage to property and the potential danger to occupants.

The right to peaceful enjoyment of one's own property against nuisance and inconvenience caused over the past five years of ownership.

Inhabitants of their own property have the right to the peaceful enjoyment. It doesn't matter who was there originally, it is the party who is making the nuisance through damage, noise, inconvenience, who is the guilty party, irrespective of the length of time the plantiffs have inhabited as against the length of time the guilty party has used the field.

59 Wilson v Kingston upon Thames Corporation (1949) 1 ALL ER 679

'A' was injured when thrown from his cycle due to the condition of the highway; a hole had been temporarily repaired (patched) with tarmacadam but it had since deteriorated again.

Held: (At court of first instance) injury caused by non-feasance rather than misfeasance in repairing the road negligently. The same decision was handed down from the Court of Appeal where Tucker L.J. stated that there was no evidence to show that the patching was negligently done.

Padbury v Holliday and Greenwood and Another (1912) 28 TLR

Mr A G Padbury was walking past a building when a tool fell from the third floor and hit his head and injured him. The building was being erected by the defendants for the building owners. The tool fell from a window sill where sub-contractors and co-defendants, Wainwright and Waring (Limited) were fixing metal casement windows. A hoarding had been put up to stop falling objects, but had been removed before the accident. At the time of the accident, the building owners had taken possession of the third floor and tenants were in occupation. The hoarding was removed by the contractor even though it was agreed the sub-contractors should use it.

This was an action for personal injuries caused by the alleged negligence of the defendants. The case raised the question of the responsibility of a contractor who employed a sub-contractor, for the injury done to a stranger through the negligence of a servant of the sub-contractor. Holliday and Greenwood claimed the negligence was on the part of the co-defendants, Wainwright and Waring only.

Held: The jury decided that there was negligence and damages of £500 were given. The Judge dismissed the contention that there was no evidence to make Holliday and Greenwood liable.

In the appeal, his Lordship said he was of the opinion that 'before a superior employer could be held liable for the negligence of a servant of a sub-contractor, it must be shown that the work which the sub-contractor was employed to do was work, the nature of which, and not merely the performance of which, cast on the superior employer, the duty of taking precautions.'

Since the tool being put on the window sill was not part of a necessary procedure to carry out the work, the negligence was on the part of a workman who was a servant of Wainwright and Waring, and not a servant of Holliday and Greenwood, who were therefore not liable for the consequences of that negligence; (it is known as collateral negligence).

The appeal was allowed.

61 **Reedie v London and NW Railway Co. (1849) 4, Exch 244**

During the building of a bridge over a highway, for the defendants, masonry fell and fatally injured the plaintiff's husband. 'R' sued on the basis of strict liability or vicarious liability on the part of the employer, the railway company.

Held: Company not liable, negligence was on the part of the contractor in the process of building, ie collateral negligence.

62   Salisbury v Woodland and Others (1970) 1 QB 324

'W' the occupier of a house employed the second defendant, an apparently competent tree surgeon, to remove a large hawthorn tree from the front garden of his house which adjoined the highway. Due to the negligence of the tree surgeon the tree when being removed fouled and broke telephone wires running across the front garden, which fell into the road causing an obstruction. 'S' a curious neighbour went into the road to remove the wires when the third defendant approached in his car driving fast and accelerating. 'S' to avoid injury threw himself on to the grass verge, but the fall caused a tumour on his spine (an angioma) to bleed resulting in paralysis.

The action is brought against the occupier claiming he is vicariously liable for the negligence of his independent contractor (the tree surgeon) when working on or near to the highway; the driver of the car is joined as third defendant. In the court of first instance all three were found equally liable and damages of £6,500 were awarded.

'S's appeal that he was not vicariously liable for the second defendant was upheld, 'he did not come within the exceptions of the general rule that each person is liable for his own torts,... the removal of the tree was...not inherently dangerous...nor was it work carried out on the highway and there was no exception in respect of work carried out near to a highway which might cause injury to persons on the highway...' The second and third defendants shared the liability and the damages; the driver being negligent in that he either deliberately ran into the wires when he saw them or in failing to see them and take evasive action.'

63   Barker v Herbert (1911) 2 KB 633

The defendant was the owner in possession of a vacant house in a street with an area which adjoined a highway. One of the rails of the area railings had been broken away by boys playing football in the street and consequently a gap had been created in the railings. The plaintiff, a child, got through this gap from the street and started clambering along inside the railings when he fell into the area and sustained injuries through the fall. In an action brought on his behalf to recover damages from the defendant in respect of his injuries, the jury found in answer to the question left to them that the area was, when the accident happened, a nuisance but the defendant did not know at the time of the accident that the rail had been removed and that such a time had not elapsed after its removal that he would have known of it at the time of the accident, and that he had used reasonable care to prevent the premises from being dangerous to persons using the highway.

See now BR v Herrington and Occupiers Liability Act 1984.

64   Wringe v Cohen (1939) 1 KB 229

'C' was the owner of a house the gable of which collapsed and damaged the adjacent property of 'W'; the pointing and collaring were defective and had been from the previous three years, the gable also had had an outward bulge. A tenant had been in occupation for two years but 'C' admitted he was liable for repairs and had always carried them out.

That the occupier is not liable for nuisance 'unless, knowing of the nuisance or being... in such a position that he ought with reasonable care to have... known of the nuisance, he allows it to continue' applies only to the particular facts, and not to cases of nuisance due to want of repair – Scrutton L. J. St Anne's Well Brewery Co v Roberts (1928) 44 TLR 703 'In our judgement if, owing to want of repair, premises on a highway become dangerous and, therefore, a nuisance and a passer-by or an adjoining owner suffers damage by their collapse, the occupier, or the owner if he has undertaken a duty of repair, is answerable whether he knew or ought to have known of the danger or not. The undertaking to repair gives the owner control of the premises and a right of access to maintain them in a safe condition. On the other hand, if the nuisance is created, not by want of repair, but a trespasser or by a secret and unobservable operation of nature, such as subsidence under or near to foundations ...neither an occupier nor an owner responsible for repair is liable, unless with knowledge or means of knowledge he allows the danger to continue...' 'C' was liable.

65   Dymond v Pearce (1972) 1 QB 496

A lorry driver parked a lorry on the highway overnight for the driver's convenience. The road was divided into two carriageways each 24 ft wide. The street lighting was on and there was a 30 mph

speed limit. The lorry, which was 7½ ft wide was parked with its lights on beneath a street lamp. It was visible from 200 yards away. At 9.45 pm a motor cyclist and the plaintiff, who was riding pillion, hit the rear of the lorry resulting in an injury to the plaintiff.
Action for damages for negligence was brought against :

(i)   the lorry driver and owner in that the parked lorry was a hazard to road users and a nuisance in that the lorry was an obstruction;
(ii)   that the lorry was parked without any right and the lorry driver as occupier of adjoining premises;
(iii)   also against the motor cyclist for negligence.

Held: Dismissing the appeal

(i)   that leaving a lorry on the highway for a considerable period for the driver's convenience constituted a nuisance by obstruction;
(ii)   that on the facts, the sole cause of the accident was the motor cyclist's negligence and the nuisance was not a cause of the plaintiff's injury; it followed that the plaintiff could not recover damages from the lorry owners or the driver.

66   **Haley v London Electricity Board (1965) AC 778 3 ALL ER 185**

The appellant, walking to work as he had done for six years, tripped over an obstacle placed by workmen for London Electricity Board near the end of a trench they were excavating in the pavement under statutory authority. The appellant was blind. The obstacle a punner hammer was placed on the pavement, so to protect pedestrians from the trench and deflect them into the road. The appellant was alone and had approached with reasonable care, waving his white stick to detect objects and obstacles. He did not detect the hammer and as a result tripped over the hammer and fell onto the pavement. The hammer gave adequate warning to normally sighted persons. The hammer was not an adequate or sufficient warning for the blind person who was taking his usual precautions by use of this stick. The appellant was entitled to recover damages at common law for negligence.

67   **Ellis v Sheffield Gas Consumer Council (1853) 2E & B 767; 23 1RQR 42**

A company which had no authority to do so instructed a contractor to excavate a trench in a street in Sheffield. The contractor's servants left a pile of excavated rubble in the roadway over which 'E' fell and sustained personal injuries.

Held: Sheffield Gas were vicariously liable for the negligence of the contractor; they had commissioned an illegal act which was followed by the negligence of the contractor ie they were liable for the consequences of their unlawful act.

68   Holliday v National Telephone Company (1899) 2 QB 392 CA

NTC were lawfully engaged in laying cables, below ground level, alongside a public street. Included with this work was the soldering or jointing of ducts which a Mr G Highmore was employed by NTC to carry out under their instructions and supervision. A flame is required to joint the ducts and whilst obtaining this flame Mr G Highmore lowered his benzoline lamp into the molten solder to warm it - this being normal procedure. However, the lamp was faulty and pressurised so when lowered into the solder it exploded injuring Holliday, a passer-by, facially as he walked along the adjoining pavement.

Negligence was not proved against Highmore for using faulty equipment whilst with the NTC working on a public highway, it was sought and found that they are required to protect the public from any likely hazard and cannot as they claimed delegate this responsibility to a sub-contractor ie Highmore.

69   Arrowsmith v Jenkins (1963) 1 QB 561

The appellant addressed a meeting from about 12.35 pm until 12.55 pm. For the first five minutes of that period, the carriage- way and pavements were completely blocked by people listening to her address. Thereafter, a passageway for vehicles was cleared by the police and a fire engine and other vehicles were guided through the crowd. The appellant, at the request of the police officers, assisted by means of a loud speaker, as a result of which the carriageway remained only partly obstructed from 12.40 until 12.55 pm, when the appellant finished speaking.

There was a finding that if it had not been for the fact that the appellant was speaking, the crowd would have dispersed and the highway would have cleared.

The accused, without lawful authority or excuse, intentionally obstructed the free passage along a highway contrary to the Highways Act 1959, S121.
If anybody by an exercise of freewill does something or omits to do

something which causes an obstruction or the continuance of an obstruction, then an offence is committed. There is no doubt that the appellant did in this case. If a highway is to be intentionally obstructed in any way they must gain lawful authority or excuse to do so, or it would contravene S121 of the Highways Act 1959.

70   Lodge Hole Colliery Co Ltd v Mayor & Corp of Wednesbury (1908) AC 323

A mine extended under a highway which was vested under S149 of the Public Health Act 1875. While working the mine the owners let down the surface of the highway which the local authority, in good faith and on the opinion of skilled advisers, restored to its former level at a great cost. An equally commodious road might have been made at less cost.

The local authority were not entitled to raise the road to the old level, cost what it might and whether it was more commodious to the public or not, and were only entitled to recover from the mine owners what it would have cost to make the equally commodious road.

The highway was vested in the local authority under the PHA 1875, but the authority were not entitled to raise the road as the new 'lower' road was equally commodious in its lower state, and the authority were only able to recover partial costs.

71   Trevett v Lee (1955) 1 ALL ER 406

In this case, the plaintiffs (Trevett) were in the process of delivering milk to the area surrounding the defendant's (Lee) house, the defendants owning the property adjacent to the highway known as Clay Lane. While completing the milk round, the plaintiff noticed the defendant had placed a garden hose on the ground crossing the highway, this being from a water tank supplied by a spring which the defendants used during months of drought, being used for only a few hours daylight. Whilst walking over the pipe the plaintiff negligently did not step clear of the hose, subsequently the plaintiff fell and sustained injuries. The plaintiff then sued the defendant for damages on the grounds of negligence and nuisance.

Per Sir Raymond Evershed made it quite clear from the start of the case there were no grounds for damages to be claimed under the heading of negligence as the plaintiff had been contributory negligent in not stepping clear of the hose. The judge then went on to refer to Salmond on Tort which conveniently summed up the to refer to Salmond on Tort which conveniently summed up the law of obstruction of highways as 'a nuisance to a highway consists either in obstructing it or in rendering it dangerous '. In the case they referred to the hose as an obstruction but then there came into play the fact that the house was adjoining the highway which led to references being made to Harper v Haden, Linke v Christchurch, Fitz v Hobson and Farrel v Mowtem, where Devlin J.'s language was referred to 'A person whose property adjoins the highway, for example, has a right of access to and from his property, and if, in the exercise of that right of access he causes as he may do sometimes an obstruction to the public using the highway, the question is whether the obstruction is reasonable or not. There are two sets of rights which have to be met and resolved on the ordinary principle that reasonable exercise of both must be allowed.' Per Sir Raymond Evershed MR concluded the case should be dismissed with the agreement of Birkett L. J. and Parker L. J.

72   **Clarke v J Sugnue & Sons (1959) The Times 29th May 1959; Digest CLY 2375**

The claim of Miss Clarke for nuisance and negligence in respect of injuries sustained when she tripped and fell on a piece of rope in the highway. His Lordship held that the defendants are not liable for the acts of independant contractors and therefore the case was dismissed.

It was shown the rope which Miss Clarke sustained injuries on did not belong to the main contractor and was, therefore, not their responsibility.

The principle of the case is that the judge ruled that a party cannot be liable for the actions of an independent contractor.

73   **Stewart v Wright (1893) 9 TLR 480**
'S' was walking along a footpath when the wind blew his mackintosh coat on to the adjacent barbed wire fence ripping his coat, 'S' claimed that the fence was a danger and a nuisance, evidence of other incidents involving torn clothing was also presented. No negligence or want of skill or care was imputed to the defendant in the erection of the fence. In the County Court (Liskeard) the fence was found to be dangerous and a nuisance and an appeal to QBD was dismissed, Mathew J. saying 'the judge came to a conclusion of fact that, this fence was dangerous and a nuisance. The principle is well illustrated by authorities...if a structure or excavation adjoining a footway is in such a condition that it is liable to do an injury, a person injured has his right of action'.

74   King (Contractors) v Page (1970) 114 SJ 355, DIG, 71/5243

A written contract between King (Contractors) and a London Borough Council enabled King to carry out some road works. In the contract, the defendants were to ensure that they or any agents of theirs were to light the road works at night.

Their agents left at the side of the road a skip which had no lights or reflectors attached to it, or even near it. At night a motorist through no fault of his own ran into the skip and injured himself.

The contractors were charged with depositing a skip on the highway without lawful authority or excuse under S140 (l) Highways Act 1959. The defendants contended that they had lawful authority or excuse under the contract they were issued to deposit the skip on the highway and that S146 (s) of the Act dealt with a failure to light things deposited on the highway.

The magistrate decided that the defendants had no lawful authority to deposit the skip and convicted them. The defendants appealed.

J. Bean said that it was in the contract that the defendants were to carry out work on certain roads. It follows that they had implied authority for all work necessary to carry out that contract, including from time to time the depositing of skips or piles of rubbish on the highway. The defendants could probably have been charged under S146 (3) for failing to light the obstruction but they had not been so charged, and no decision on that point was called for.

The conviction under S140 (l) Highways Act 1959 was wrong and should be quashed.

75   Gatland v Metropolitan Police (1968) 2 QB 279

A skip was deposited on a public highway with flashing lights as a warning, but there was no permission from the local highways authority. A lorry collided with the skip in question causing no injury to the driver but damage to the vehicle.

The contractors (who placed the skip) were convicted under S140 (l) of the Higthways Act 1959, that the skip had been deposited unlawfully.

The principle of law is that nothing should be deposited on a public highway that could endanger other users of the highway.

76   Hunston v Last (1965) 109 SJ 391

The defendant was burning stubble in a field well beyond the legal distance (50 ft) from the highway. The fire spread to a grass verge and then downwind to the highway causing smoke and flames to obstruct the highway. Two cars then collided on the highway. The defendant was charged with 'without lawful authority/excuse lighting a fire within 50 ft of centre of highway'. In consequence, passage of the highway was interrupted. Although fire was lit more than 50 ft from highway, spread was inevitable. The defendant was convicted and fined £2. He appealed. However, the defendant had conceded that a person could be said to have lit a fire within 50 ft of highway if he lit it beyond 50 ft with the intention of the fire spreading. Therefore, for this case, a man who lit a fire committed an offence since the justices had approached the matter in a way which saw the spread of fire as inevitable, due to evidence showing this. Appeal dismissed.

77   Farrell v Mowlem (1954) IL Lloyds Rep 437

Farrell, the plaintiff, sustained personal injuries when he tripped over a compressor pipe while walking along the pavement with his friends. The compressor was located 50 yards along the street to reduce noise level and disperse the plant. The pipe was run from the compressor, parked on the road across the pavement into the front gardens of houses and back across the pavement work area. The plaintiff alleged negligence by permitting the pipe to be laid out in a dangerous and improper manner, he also or alternatively alleged nuisance to the highway. Hence this case is a test of whether a member of the public using the highway is entitled to expect it to be free from nuisances. The defendants, John Mowlem & Co Ltd, alleged contributory negligence accusing Farrell of being thoughtless in not associating workmen and plant with hazards and also accused him of being careless in not seeing/expecting the second pipe after he successfully crossed the first one. The contractors were working for the council and therefore in the public interest and were within the law.
The Judgement: It was not really necessary for the pipe to cross the pavement twice since it could have been put in the gutter or the compressor moved closer to the plant nor is it reasonable to expect people to cross the road or expect a nuisance when walking in a group in a London street.
Held: Nuisance on the highway. John Mowlem ordered to pay £87 damages to Farrell.

Note that Harper v G N Haden & Sons Ltd (1933) is *not* the true principle in *this* case.

Even if you are acting within the law on the roadside you are not entitled to cause a nuisance to pedestrians using the pavement unless there is a valid reason and then adequate warning must be given to users of the highway.

If a contractor causes a nuisance he could be liable for any injuries or inconvenience arising from that nuisance that the public encounters.
(i) contractor can only lay pipes, machinery etc on the pavement if it is the only way work can proceed and
(ii) then the contractor must warn the public of this danger - using signs and instructing workers of this responsibility;
(iii) the compressor will be located next to the work site, thus eliminating pipe length, and increasing noise level;
(iv) the contractor should run the pipe in the gutter and accept the risk of damage to the pipe, however if someone trips over the pipe in the gutter, the contractor may still be liable.

78   **Carshalton UDC v Burrage (1911) 2 Ch 133**

An ancient unfenced chalk pit separated from the highway by a strip of land varying in width (14 to 50 ft, 295 ft long) 'B' owned the land but not the pit. With erosion caused by the weather, the side of the pit had moved into the strip of land. Public Health (Amendment) Act 1907 empowered to fence, or require owner to enclose, 'in any situation fronting, adjoining or abutting any street... any excavation or bank'.

Held: The pit was as described above and the L.A. had power to require 'B' to erect a fence on his land enclosing the pit.

79   **Myers v Harrow Corporation (1962) 2 QB 442.**

In the Spring of 1959 the appellant bought a house and land from builders who retained adjoining land. In August 1959, the builders entered into an agreement with local highway authority under S40, Highways Act 1959, for the construction of a road. Between 1959 and 1961 the road and footpath was constructed. The footpath ran part of its length alongside the appellant's premises. The path was level at one end with the lawn and raised 13 inches at the other end; the total fall with kerbstone was 15½ inches. In the Spring of 1959 the appellant's lawn was the same level as the land on which the road had been constructed, and its level had not been altered.

The Highway Authority served notice on Myers. Myers appealed against the notice, but justices holding that the appellant's lawn was in fact dangerous to persons using the road.

Held on appeal: The words in Highways Act 1959 S144 (i) 'Source of danger' and 'in or on any land' were descriptive of a danger in or on the land to which the user of the highway might be subjected by something escaping or coming off the land, or which might be encountered if he inadvertently stepped off the highway; and that, although the difference of levels, meant there was a danger of someone inadvertently stopping off the road, it was not 'in or on' the appellant's land. There was therefore no unfenced 'source of danger' within the meaning of S144 (i) in or on the appellant's land and the Justices decision was wrong.

80 **Nicholson v Southern Railway Co & Cheam UDC (1935)
1 KB 558; ALL ER 168**

Miss Nicholson was walking along Station Road, Cheam, Surrey at night and fell from the highway onto the adjoining land which was owned by the first defendant, the Southern Rail Co. The land was unfenced and Miss Nicholson broke her leg due to the fall.

The plaintiff contended that the second defendant, Cheam UDC, the highway authority, had made up the road and in doing so had created a 6 inch drop between the highway and the area adjoining. The Cheam UDC had allowed this unsafe condition to remain.

The first defendant, the Southern Rail Co, denied that they had been negligent and as they were not occupiers of the land they were not responsible for its condition.

The second defendant, the Cheam UDC denied that they had committed a breach of duty to the plaintiff or were guilty of any negligence. They pleaded that the landowner was guilty and that the plaintiff was guilty of contributory negligence.

When works have been done on a highway which render it dangerous, there is no obligation on the owner of the adjoining land to fence his land or alter it in any way to remove the danger.

His only obligation is not to allow his land to be used in such a manner as to constitute a nuisance to the highway or persons using it.

Held: Plaintiff entitled to recover £500 damages from the Highway Authority. Case against the Southern Railway Co. dismissed.

81   **New Towns Commission v Hemel Hempstead Corporation (1962) 3 ALL ER**

A large office building on each side of a highway and extending across the highway was proposed to be erected. The span across the highway would be formed by the three top floors of the five proposed floors of the building. The space below the span would be empty, except for three supporting columns about 17 ft high, the span would extend along and over the highway for about 43 ft. The highway could be crossed by walking along a corridor in the building. It was questioned whether a licence or consent would be needed under sections 151 and 152 of the Highways Act 1959 for the construction of the building across the highway.

It was decided that it was not the primary object of the building to obtain a passage across the highway, so neither section 151 nor 152 apply to this case.

82   **Pitting v Abergele UDC (1950) IKB 636**

The case was brought by Pitting because the local authority refused a licence for a mobile dwelling on a site. The licence was refused because 'such would be detrimental to the amenities of the district, particularly on account of the proximity of other dwellings'.

Pitting appealed against this decision to the Justices contending that the grounds given were outside the scope of the Public Health Act 1936.

The case was dismissed on the grounds of S269 of the Act under which it empowers local authorities to refuse such a licence. Pitting appealed against this, contending that the local authority had really refused the licence on town planning reasons, and not as stated on public health grounds. The result of the appeal was that Pitting was granted a licence on the grounds that the authority's reasons for refusing the licence were the wrong ones.

83   **Principality Building Society v Cardiff Corporation (1968) P & CR 821**

The plaintiff (Principality Building Society) owned a piece of land consisting of a basement containing cellars, warehouses and showroom. The defendant (Local Authority) built a road over this cellar. Under a conveyance dated 10 November 1914, it was the responsibility of the plaintiff to maintain the roof of the cellar of sufficient strength to carry the traffic of the district. There were three traffic islands in the road which were formed as part of the cellar roof making them the plaintiff's property, these islands allowed daylight to enter the cellar through roof lights. Traffic conditions changed in recent times and the cellar could no longer withstand the loads of present traffic. A dispute arose as to the plaintiff's liability under the 1914 conveyance.

It was held that the conveyance referred to the building at the time of the conveyance and the traffic conditions at that time, and the plaintiff's obligation was to maintain it in its original state. It was also held that the defendants were not entitled to interfere with the traffic islands without permission of the plaintiff.

The principles of law involved were mainly the liability under a conveyance and the interfering with a structure which does not belong to the interfering party.

84   Drury v Camden LBC (1972) RTR 391

Drury at 4.55 am on 4 March 1966 collided with a builder's skip in Camden Street, NW London. The council were, and had been for some months, carrying out building works using hired builder's skips, for disposing of rubble in the street. There were three lanes which were 35 ft wide, the skip obstructing 6 ft of the road beside the kerb. The skip was unlit and the same colour as the road surface. The plaintiff knew the road well and said he had not seen the skip until he was 15 ft from it. The Judge said that the skip was a nuisance and the council were negligent for not lighting it, but the plaintiff was found guilty of contributory negligence, and the Judge accordingly awarded the plaintiff half the damages that he would (£3,590.46) from the council.

85   Hardcastle v Bielby (1892) 1QB 709

The case was stated by justice for the West Riding of Yorkshire, on the appeal of John Hardcastle, surveyor of highways to the Hodworth Local Board, against a conviction for causing a heap of stone to be laid upon a highway and allowing the same stone to remain there at night contrary to the Highways Act (5 & 6 wm 4c50) S56.

On 4 November 1891 a carter acting under the orders of a person to whom the appellant had given general directions as to repairing the highway in question, placed two loads of stone in heaps on the road and at about 7pm that evening a cart collided with the stone and was injured. It was admitted that the carter who placed the stone upon the road had seen a paper signed by the appellant (his foreman) containing a direction that the materials were not to be placed 'so as to cause obstruction of the road'. The question which the case was based on was whether or not a surveyor could be convicted under the Section, 'apart from personal knowledge of the act of his 'foreman' in directing the stone to be laid on the road and allowing it to remain there at night'. Mr R M Bray for the appellant, the conviction cannot be supported. The language of the section in no way suggests that the statutory misdemeanor can be committed without a guilty intention. There is no evidence that the surveyor either 'caused the heap to be laid' on the highway or 'allowed it to remain there that night'. The case showed that he did not even know that it had been left on the highway at all.

86   **Saper v Hurgate Builders Ltd and Others, King v Hurgate Builders Ltd and Others (1971) RTR 38 DC**

'HB' were carrying out work at 'B's house. At 'HB's request 'R' a self-employed joiner working on the job for them, obtained at their request a skip from 'SG(H)' who placed it partly on the footpath and partly in the road, it was 60 yards from the blind crest of a hill. 'B' at the request of 'R' placed lighted lamps on top of the skip at each offside corner. At about 12.30 am 'K' in his E type Jaguar with 'Miss S' as passenger collided with the skip causing injury to both of them. The Jaguar and the skip were damaged. The injured parties alleged negligence and breach of statutory duty under S146 of the 1959 Act against 'HB' and 'SG(H)', actions heard together.

Held: Nuisance and negligence on the part of 'HB' and 'SG(H)' in relation to 'Miss S', and 'K' also negligence on the part of 'K' to 'Miss S' 'an alert motorist would have seen the skip in time to avoid it... 'K' was guilty of contributory negligence to 40%'.

Per Cuniam: S146 '59 Act (now S80 Act) does not provide a cause of action for an individual injured by a breach of it. (HB(H)' having left the skip in the road were liable to the plaintiffs; they could not escape liability to third parties by delegating the lighting of the skip to someone else. 'HB' were liable for having caused the skip to be placed in the roadway and 'R' who was a servant of 'HB' (interesting distinction made between contract of service and contract for services not language used) was equally liable; 'B' not having failed to do properly what he had been asked to do, no judgement against him (by concession). 'HB', 'R' and 'SG(H)' should bear 50% and 'HB' and 'R' 50% jointly; but that 'R' is entitled to be indemnified by 'HB' since he had done previously as instructed. 'HB' as bailers of the skip, failed to take care of it and were liable to 'SG(H)' for the damage caused to it by the Jaguar crashing into it.'

87   **Lambeth BC v Saunders Transport (1974) RTR 319 DC**

'S' the owner of builder's skips hired one out to a hirer on
conditions set out on the reverse side of a printed form. On the face
of the form it stated that the hirer was responsible for lighting the
skip according to the permit granted by the local authority which
required lamps at night, lamps were available for hire from 'S'. No
lamps were used and 'S' was prosecuted for failing to ensure that
the skip was lighted contrary to S31(4) ('71 Act). 'S' raised defence
of 'another person' and that 'S' had taken all reasonable
precautions and exercised all due diligence to avoid the offence. At
both magistrates court and the Divisional Court it was held that 'S'
had made out the defence.

88   **Barnet LBC v S and W Transport Ltd (1975) RTR 211**

The owners of a builder's skip had permission from the highway
authority for the skip to be deposited on a highway. They hired the
skip to a contractor, placing on him a contractual duty to light it. It
was on the highway but was unlighted on several nights because,
as the authority knew, the lamps which were on the skip were
unlit. The authority charged the owners with offences of
contravening Section 31 (4) (a) of the Highways Act 1971 in not
securing that the skip was properly lighted during the hours of
darkness. The owners gave to the authority notice under Section
31(7) of intention to rely on the defence in Section 31(6), the notice
stating the contractor's name and contractual duty and giving as
the reason for the lack of lighting, as the owners then believed, that
the lamps had been stolen by an unknown thief. At the hearing the
defendants for their defence relied on the act or default of the
contractor, who accepted responsibility for lighting the skip. The
authority raised objection that the notice under Section 31(7) was
defective and that the hearing should be adjourned for the owners
to give an amended notice. The justices proceeded with the hearing
and were of opinion that, although the authority had proved the
commission of the offence, the owners had established the
statutory defence under Section 31(6), and the informations were
dismissed. The authority appealed.

Held: Dismissing the appeal, that on the true construction of
Section 31(7) of the Highways Act 1971, the obligation on the person
charged with an offence under Section 31, who desired to give the
notice preparatory to putting forward the statutory defence under
Section 31(6), was to give information as full as he was able to
provide and in accordance with the facts then in the possession
that the justices had jurisdiction to adjourn the hearing for an

amended notice but, in the circumstances, were not obliged or required to do so because the authority by that time had all the necessary information which Section 31(7) intended the owners to provide.

89   York City Council v Poller (1976) RTR 37 DC

The owner of a builder's skip hired from him by the Council for about 4 weeks, was told by the housing department to put the skip on a highway. On asking for permission to do this under Section 31 of the 1971 Highway Act, he was told by a council official that a blanket permit was in force authorising skips to be deposited anywhere on the Highway directed by the Housing Department. No written permission in accordance with Section 230 of the 1959 Highways Act was given by the council and no such permit as a 'blanket permit' was in force. The owner deposited the skip on the highway as directed and a motor cyclist was fatally injured on colliding with the skip at night. The council were charged under Section 31(5) of the 1971 Act of depositing the skip on the highway without permission contrary to Section 31(3).

The justices were of the opinion that the offence committed by the owner was due to the act or default of the council because the council who ought to have known about such measures had said that there was a blanket permission when in fact there was not and the owner committed the offence in reliance of the statement. The council were convicted.

90   Pitman v Southern Electricity Board (1978) 3 ALL ER 901

At the end of their day's work the defendants' employees left a metal plate, standing about one-eighth of an inch proud of the surrounding pavement, over a hole in which they were to place a junction box. The hole was not marked by any light and, when it was dusk, an elderly local inhabitant tripped over the metal plate and fractured her wrist.
Held: By altering the condition and level of the pavement the metal plate had introduced an unexpected hazard which constituted a potential danger to users of the pavement and the county court judge had therefore been entitled to find that the defendants had caused the lady's injury by their negligence.

91   Wills v T F Martin (Roof Contractors) Ltd (1972) RTR 368

At about 6.00 am the plaintiff was pedalling his moped down a dimly lit residential road in order to get the engine started and the lights on. The defendant had deposited a skip in the roadway close to the kerb on the plaintiff's nearside. The plaintiff had failed to see the skip and collided with it resulting in injuries to himself. It was claimed that the placing of a skip was not a proper use of the highway and so caused a nuisance regardless of it being lit or unlit.

However, the plaintiff revealed that when he was on top of the hill, he could see dimly the outlines of cars beyond the skip. Placing of the skip was a nuisance, but the negligence of the plaintiff in not looking where he was going was greater than the negligence of the defendant who had not provided *adequate* lighting and was the cause of the accident. (Adequate lighting takes into account street lighting and surrounding trees etc.)

(i) Defendant was negligent in leaving the skip on the highway without warning lights thereby endangering the safety of other road users.
(ii) Depositing the skip in the road constituted a nuisance.

Held: plaintiff had not been able to prove that the skip was lit and it was his failure to look where he was going that was the sole cause of the accident. The plaintiff's negligence was greater than that of the defendant. Dymond v Pearce (1972) RTR 169 - case involving a parked lorry on a dual carriageway which was hit by a motor cyclist.

92   **Derrick v Cornhill (1970) Crim LR 467, CA**

The defendant hired a hopper from a hiring firm on the terms that he would obtain the necessary permit for the hopper to be placed on the highway. He also said that he would see to the lighting of the hopper. Unfortunately the defendant forgot to acquire a permit and also to light the hopper. Because of this mistake he was charged with depositing the hopper unlawfully under the Highways Act 1959 ss 127(b) and 140 (i). On his behalf it was contended that he was not present when the hopper was delivered, and did not sign the delivery documents, and did not direct its siting.

The defendant's appeal against the conviction was dismissed as the hopper had been delivered by the driver of the hiring company on the defendant's instructions and, so, in law, the deposit by the driver of the hopper was a deposit by the defendant.

On 4 September 1970, it was alleged that the defendants, C Gabriel Ltd, placed a metal skip outside a house in Lavendar Hill, Enfield, and did without lawful authority or excuse obstruct the free passage of the highway by thus placing the skip, contrary to Section 121 of the Highways Act, 1959. The driver delivering the skip produced a contract of hire which the customer signed, then asked the customer where to place the skip. The customer requested that the skip be placed outside the house on the highway. The defendant's form of contract incorporated the standard conditions of hire; one of which was 'the customer undertakes to direct at his sole discretion the driver... where to deposit the skip, the said driver being for the purpose of such deposit the agent of the customer'.

One of the Council's highway officials saw the skip standing there and requested the defendants to remove it, which they did later the same day. No authority was given by the Council to leave the skip there, and the defendants were taken to court. The Justices were of the opinion that although the driver was acting on the instructions of the customer, he was still acting as the defendant's employee, and that the contract did not relieve them of any criminal liability. The defendants were convicted and fined. They appealed. Giving judgement on the appeal, Lord Justice Widgery said that as the offence under the Highways Act was 'absolute' the driver who had deposited the skip was guilty of an offence for which his employers (the plant hirers) were liable as they had necessarily delegated to the driver the final decision as to where the skip was to be placed. Criminal responsibility could not be avoided merely by agreeing with another person that the criminal act was to be regarded as his; therefore, the conditions in the contract of hire did not alter the situation and did not give the hire company a good defence against a charge of having obstructed the highway. His lordship added, however, that as the customer had authorised, permitted or encouraged the deposit of the skip on the highway he too could have been found guilty of an offence. In this connection, his lordship referred to the earlier case of Derrick v Cornhill, in which the contract and not the hire company was convicted of an offence after the illegal depositing of a hopper on the highway. The legal position, as outlined by his lordship, was that in a case such as the one under consideration the driver was acting not only for his employers but also for the contractors who had hired the skip, so that both had a legal liability.

Hales Containers placed a skip on the highway, Hales pointed out to the customer the necessity for lights at night and that it was a customer's responsibility.

The skip was 10 yards from a street light and the road was 25 ft wide at that point. Lights were not put round and a Mr Jones's car collided with the skip at approximately 6.50 pm on the 26 October 1968. There was damage caused to the car but no personal injury was sustained. The case was a prosecution (S146(3) '59 Act) before magistrates against 'Hales' for failure to light the skip: Hales was convicted. 'H' claimed that the responsibility had been transferred to the customer.

Held:
(i)   the provision to 'light' was strict liability on the part of the owner;
(ii) the obligation had not been transferred.

95   **Westminster Bank v Beverley BC (1968) 2 WLR**
       1080(1969) 2 ALL ER 104; (1969) 1 QB 499 2 ALL ER

Westminster Bank applied for planning permission to build an extension to the rear of a branch that they owned, but were refused permission by Beverley Borough Council, who were the local planning authority. After a local enquiry, which was later re-opened, the Bank's appeal to the Minister was dismissed. The reason for this dismissal of appeal and planning permission was because of the Local Planning Authority's proposals for a town centre development. This development entailed a road widening scheme to the rear of the bank, although no prescribed date was given. This effectively meant that the Bank could not build its proposed extension. This road widening scheme was approved by Beverley Borough Council, acting as the Highway and Planning Authority, this proposal being a firm one.

At the re-opened enquiry, Beverley Borough Council said that in recent years any proposal for development within their proposed 'improvement line, had been resisted, which in effect meant that an improvement line had been proposed officially. No improvement line had however been prescribed under the relevant statutory power for the road widening of the Highways Act (1959). Under this Act, any party who is injuriously affected by the prescribing of an improvement line, is entitled to compensation. The Borough Council realised that a refusal of Planning Permission would deny compensation to the Bank. On the Bank's application to a higher court for the records of the case (certiorari) to quash the Minister's decision, the court found that the real reason for the refusal of planning permission was because of the compensation involved. Because the Bank was injuriously affected by the improvement line, it was entitled to compensation, but the Borough Council tried to take the cheap way out by refusing planning permission, thus incurring no compensation claim. (Affirmed in (1971) AC 5030.)

96 Sittingbourne LBC v Liptons Ltd (1931) 1 KB 539

In 1908 Liptons became the tenants of 106 to 108 High Street, Sittingbourne, and with the approval of the appellants (Sittingbourne LBC), set back the ground floor frontage.

In 1925 the appellants adopted Part II of the Public Health Act S3. This stated 'No new building erection or excavation shall after an improvement line has been prescribed be placed or made nearer to the Centre Line of the street than the improvement line except with the consent of the Local Authority. This was in order that if need be the road or street may be widened. In June 1927 the appellants duly set an improvement line in the High Street.

On November 15 1929 work was started on the shop of which the respondents had not submitted plans or applications for consent. The work consisted of moving the shop frontage back to its original position which due to the imposed improvement line went over the improvement line, the work involved putting up a stud partition, door and a wooden shop front faced with marble. This work did not constitute a load bearing structure. The appellants contended that this work was an erection under S33 sub s(1) of the Act 1925.

Whether or not the work carried out was an erection as stated in S33 sub s(1) of the Public Health Act 1925.

Held: The justices contended that the work did not constitute an erection and dismissed the case.

97 Robinson v The Local Board for the District of Barton-Eccles, Winton and Monton (1883) 8 AC 798

G Robinson, the appellant, intended to build a row of nine houses, fronting New Lane, Patricroft. The local board 'disapproved' of his intention (as a written description) 'on the ground that the houses will contravene the building line'. The local board had said that they would remove the buildings as laid down in the Bye-laws created, as allowed by the Public Health Act 1875. The appellants claim that not only was the street an old, rather than new, street, but also that the local board could not disapprove because the houses crossed the building line. It was claimed that the street was not a 'new street' because of the existing houses there, and stated by Lord Blackburn that it 'was an ancient highway which came within the definition of 'street' in the interpretation clause of the Act of 1848'.

The court decided that under the existing bye-laws, they could not disapprove of and pull down houses because the building line was too near the roadway.

98   **Devonport Corporation v Tozer (1902) 2 Ch 182**

The defendants were the ownners of a triangular piece of land within the plaintiff's borough. Two sides of the triangle abutted upon public highways within the borough. The defendants, in pursuance of a building scheme, erected houses on their land fronting the highways. The plaintiffs alleged that the defendants were laying out the highways as new streets which did not comply with the requirements of the borough bye-laws as to width, and they claimed, first, an injunction, and secondly, a declaration that the plaintiffs were entitled to remove or pull down any work begun or done by the defendants in contravention of the bye-laws. The bye-laws, which were framed under the Public Health Act 1875, prescribed and provided that the plaintiffs might, subject to any statutory provision in that behalf, remove, alter, or pull down any work begun or done in contravention of the bye-laws.

Held: That as the adjacent owner has access to the highway from his property. No trespass upon the highway was committed.

(1) That the defendants were not laying out or intending to lay out the highways as new streets within the meaning of the bye-laws, (ii) that the bye-laws could not be enforced by action for an injunction, but only by the special remedies, provided or by way of information by the Attorney-General (iii) that no such declaration as asked for ought to be made.

99   **Astor v Fulham BC (1963) 61LGR 281**

'A' owned sports fields on either side of a road, about half a mile long, which had been a public highway since time immemorial. Prior to 1855 there was a mansion on one side and a school and some alms-houses on the other. Other buildings were erected by 1930, but at that date half the length of the road on one side and most of the other were bounded by sports fields. In 1951 the LCC compulsorily purchased some of the land and built a school, seven blocks of flats and converted a house into a hospital. In 1959 FBC passed a resolution declaring the road a new street and apportioning cost on a frontage basis. 'A' challenged this claiming: the road was not a 'new street' under the 1855 Metropolis Management Act, before the Act it was not a street in the ordinary sense but was converted into one some time later (following developments) and became a 'new street' for the purpose of S105 of the 1855 Act and that it ceased to be a new street after a lapse of time. Held: dismissing summons; a 'new street' for purpose of Act was a matter of fact, and Sullivan Road became one in 1951 when its character was changed; it possessed the characteristics of a street and was a 'new street' within S105.

100   **Buckinghamshire CC v Trigg (1963) 1 ALL ER 403**

'T' lived in and owned an upper maisonette; when the street was 'made up' he was served with notice regarding payment of streetage, the council claiming that 'T's property adjoined the street. Council upheld by magistrates court; this case is an appeal by way of case stated to the Divisional Court of the QB (three judges sitting).

Held: The words 'front', 'adjoins', 'abuts' in 1959 Act (similar in 1980 Act) envisages actual contact of the sort that will produce some frontage which can be measured. The provisional apportionment was wrong because on the facts it was impossible to say that 'T's upper maisonette adjoined the street since it was separated from it by the garden in the ownership of the respondent.

101   **Warwickshire CC v Adkins and Others (1968) 66LGR 486**

When developing a housing site adjoining a private street a strip of land 12 feet wide was left at the boundary as a planning condition to allow for future road widening. The housing site and the strip were in different ownership: the strip was uncultivated and was used for such things as car parking, the adjoining properties having an easement over it to obtain access to the private street. The action arose out of the decision of the local street works authority to make up the street and allocate the cost on a frontage basis. The householders claimed that they did not front the street and so therefore were not liable to pay streetage costs.

The magistrate found that there was no evidence that the strip had been absorbed into the private street and the 'resolution of approval' was quashed. The court allowed the appeal from the authority stating that the only proper conclusion was that the strip had become part of the private street 'the justices must have taken into consideration matters which were not entitled to be considered, or must have misdirected themselves in law'. (No question of dedication is involved.)

102   **Ware UDC v Gaunt (1960) 3 ALL ER 778**

A new road was laid out which ran alongside an existing public footpath (repairable at public expense and therefore not within the definition of 'street' included in S5 of the Private Street Works Act 1892) which fronted a school. The LA in 1956 resolved as required by the above Act to make up this road and to apportion costs to the frontages including the school. The specification which was approved by the LA following the resolution referred to the provision of '...street lighting...surface water drainage, gulley posts...' 'G' challenged the LA ultimately to the QBD.

Held:
(i)   this road was within the definition of street;

(ii)   unless '...street lighting, surface drainage' etc were included in the resolution they could not be included in the specification of the works and should be struck out; and

(iii) although the school was separated from the road by the public footpath (the argument is that only frontages which adjoin the road pay) the school did adjoin the road within the words 'fronting, adjoining or abutting' and therefore could be charged.

103   **Littler v Liverpool Corporation (1968) 2 ALL ER 343**

'L', a 19 year old youth who had the handicap of constitutional backwardness, was running to the corner shop to return lemonade bottles and to buy new full ones. He tripped over a defective York stone paving slab, fell, and the bottles he was carrying smashed, the broken glass causing him personal injury which is the reason for the action. A corner of the stone flag had flaked off leaving a triangular shaped hole 3 inches (75 mm) long on its longest side and about half an inch deep (12 mm). The claim is that the reason for the accident was the unevenness of the pavement and the flaked corner. There were slight irregularities, the pavement was about 80 years old, and was in constant use by men, women and children. 'L' claimed failure to maintain highway in safe condition or rope off a dangerous part or give adequate warning.

Held: LA not liable. 'Uneven surfaces and differences in level between flatstones of about an inch may cause a pedestrian temporarily off balance to trip or stumble, but such characteristics have to be accepted. A highway is not be criticised by the standards of a bowling green...' per Cumming-Bruce J. at p.345.

104 **Sinclair-Lockhart's Trustees v Central Land Board (1951) P & C R (Ct of Sess).**

Following S66 of the Town and Country Planning (Scotland) Act 1947 on application for development the Central Land Board fixes for each development a development land charge. 'S' wished to convert an existing farmhouse into two cottages and build a new seven roomed farmhouse. A charge of £20 for the conversion and £140 for the new house was made. 'S' appealed against the £140, claiming that the board should have regard only to the land enclosed within the outside walls of the proposed building when assessing the charge to be imposed. The board replied (i) that the action was incompetent as the Act placed on the board alone the duty to determine what if any charge and (ii) that the action was irrelevant as the S67 (a) meant such area of an applicant's land as might reasonably be regarded as being improved in value. The decision was in favour of the board per Lord Mackintosh:

'Ground which is used for the comfortable enjoyment of a house or other building may be regarded in law as being within the curtilage of that house or building and thereby as an integral part of the same, although it is not marked off or enclosed in any way. It is enough that it serves the purpose of the house or building in some necessary or reasonably useful way.'

105 **Whitstable UDC v Campbell (1959) QB JPL 46 (1959)**

In early 1957, Mrs Campbell ('C') instructed architects to design a new building on her land. The plans were sent to the council, but were rejected being contrary to the New Streets Act 1957 (1). However Mrs C commenced building – Whitstable UDC took her to court.

It was revealed in court that in May 1957 notice of refusal was sent by 'W' addressed to 'C' at the address of her architects and not to her home address. Further correspondence was subsequently made by letter and 'C' had a telephone conversation with a member of the council, where she was warned of the consequences of commencement of her building project before complying with the 1957 Act.

In her defence 'C' contended that her notice had not been served in the manner prescribed by the 1957 Act (S8) and by the Public Health Act 1936 (S285), because it had been addressed to her at her architects' address and not at her home address. Failure to serve notice in the prescribed manner was a fundamental defect which made the notice void.

'C' won her case and was acquitted.

However, 'W' appealed, contending that even though the notice had not been sent in accordance with the Act, it was sent to the address from which the plans were received, ie the architects, which was the normal procedure. Also that the ensuing correspondence with 'C' showed that the notice had been received and acknowledged thus curing any defect.

The fact that notice was not sent in the prescribed manner is irrelevant if the notice is received, as it was – only important if recipient does not receive notice.

Held: In the Chancery Division that if the notice is acknowledged in this case there was no dispute of that, then 'C' must comply with that – 'C' fined £15.

106 National Employers' Mutual General Insurance Association v Herne Bay UDC (1972) 70 LQR 592

Under a S40 1959 Act agreement (S38 1980 Act) the builder agrees to construct the roads and sewers at his own expense under the supervision of the highway authority. To cover for the possibility of him not completing the work the authority obtains a surety which will safeguard the completion of the work. On this particular job the builders went into liquidation with the roads only partially completed. The remaining plots were sold to other builders on the understanding that there would be no street charges. The case was brought by the plaintiffs to determine the limits of their surety liability, they claimed that the new builders should pay 'their share'.

Held: The surety crystallised on the liquidation of the original builder and subsequent builders were under no obligation to pay street charges (this case is also valuable regarding the interpretation of a contract).

107   Griffiths v Liverpool Corporation (1966) 2 ALL ER 1015

A pedestrian slipped and suffered injury when walking on a flagstone which protruded half an inch above the rest of the pavement. It was found as a fact that the flagstone was dangerous. On the evidence a regular system of inspection was desirable, and labourers could have been found to make such an inspection, but if dangers had been discovered there were insufficient workmen to make them safe. The particular flagstone could not have been made safe.

Held: The plaintiff's claim should succeed. The highway authority had not proved that they had taken such care as was in all the circumstances reasonably required to make sure the highway was safe. Liability for breach of statutory duty for non-repair of the highway under S1 of the Highways (Miscellaneous Provisions) Act 1961 is absolute, subject to the statutory defences made available by the Act (Now S58 'BD' Act).

108   Rider v Rider and Another (1973) 2 WLR 190

The plaintiff was injured when a car driven by the defendant (her husband) swung across the road and collided with an oncoming van, the driver of the van was not to blame for the accident. The road was a rather narrow unclassified one which had become a 'secondary through route' between two centres of population; it had no real foundations and the edges were crumbling; the defendant 'R' was familiar with the condition of the road. The court of first instance found that the 'swing' of the car was probably caused by a wheel hitting an obstruction and 'R' losing control of the steering. The road because of its condition was foreseeably dangerous to reasonable users and the standard of repair maintained by the highway authority (Another) had been inadequate in the circumstances but because 'R' (the driver) knowing its condition should have been driving more slowly he was one third liable and the authority two thirds so.

109   Hardaker v Idle District Council (1896) 1 QB 335

The defendant council employed a contractor to construct a sewer. Due to the negligence of the contractor a gas pipe was fractured and the escaping gas exploded damaging the plaintiff's property. It was held that the defendants could not relieve themselves from the breach of duty to take care by employing a contractor. Accordingly, they were held liable towards the plaintiff.

110 **Burnside and Another v Emerson and Others (1968) 3 ALL ER 741**

'B' driving his care at 25 miles an hour in pouring rain was in collision with 'E's car which was travelling in the opposite direction and which, whilst passing through a pool of water on the partially flooded highway (halfway across) resulted in 'E's car coming across the road and colliding with 'B's car. 'E' was killed, 'B' and his wife were injured, and subsequently damages they were to receive were agreed with 'E's estate; they also sued 'E's estate claiming that 'E' 'pulled out' and was at fault. 'E's estate joined Nottingham CC, the local highway authority, claiming that they had not fulfilled their duty of adequately draining the highway. At court of first instance the authority were held fully liable (therefore 'B' will receive the money, the CA appeal is now related to who pays 'E' or 'NCC').

Appeal allowed in part. The LHA were negligent but the driver was driving too fast for the conditions and the LHA contribution was one third (for negligent maintenance) and 'E's was two thirds.

111 **Haydon v Kent CC (1978) 2 ALL ER 97**

'H' was injured when she slipped on ice which had formed on a steep footpath adjacent to her home, this occurred on 5 February 1973, the icy conditions arose on the previous Sunday or Monday, and the path was slippy by Tuesday – on Wednesday a local authority workman used the path and reported its bad condition; 'H' had her accident on the Thursday, before the authority gritters arrived. 'H' brought an action for breach of duty to maintain the footpath under S44 (i) of the 1959 Act and claimed damages. The LA did not claim the statutory defence (Sl(2) 1961 Act). Court of the first instance held that the authority were in breach of duty by not clearing the snow and ice earlier.

In the Court of Appeal the appeal was allowed on the basis of: 'the duty to maintain...embraced only the duty to repair...and did not include a duty to remove obstructions, except where damage to the surface of the highway occurs, a transient obstruction is not of this kind' (per L. Denney) 'the duty can have wider scope than a duty to repair and could include a duty to remove obstructions, but 'H' could not establish breach of that duty as following notification the authority had acted promptly' (per Goff and Shaw L.JJ.)

112   Meggs v Liverpool Corporation (1968) 1 ALL ER 1137

The plaintiff, an old lady of 74, tripped over an uneven flagstone on the pavement and was injured. The pavement was in continual use, but the defendants had received no complaints about its state.

The plaintiff claimed damages.
Held: users of the highway must take account of the possibility of unevenness in the pavement, and the evidence was not sufficient to establish a prima facie case that the pavement was dangerous to traffic or that there was a breach of obligation to keep the pavement in repair. Consequently, the plaintiff's action failed.

113   Burton v West Suffolk CC (1960) 2 ALL ER 226

Drainage work done on the roadway by the authority was properly done but was inadequate. 'B' was injured when his car skidded on ice which formed from water lying on the road as a result of the poor drainage. When the road flooded the authority put out signs but when icy, did not. 'B' claimed misfeasance on the part of the authority for the poor drainage and also that signs should have been put out.

Held: The drainage work whilst inadequate was properly carried out and was not misfeasance; also there is no duty to warn of icy conditions.

114   PGM Building Co v Kensington and Chelsea Royal LBC (1982) RTR 107 DC

In the course of business PGM hired a skip which with the permission of the Highway Authority was placed on the highway. On many occasions during the hours of darkness the lights placed around the skip were stolen or broken. From time to time PMG would replace the lights, but there were times when the skip was left unlit. The authority laid an information against PMG under the Highway Act 1971. PMG notified the authority that they were going to rely on the defence of '...another person...' S31 (6) and (7) of the 1971 Act, but which calls for the identity of the culprit to be known, that is the defence requires a name; PMG did not have one, the culprit was unknown. The magistrates held that without a name the defence did not apply. In the Divisional Court it was decided that it was a defence to prove the offence had been committed due to the act or default of an unidentified person and that a notice served under S31(7) did not have to specify an identifiable person.

115   **Whitaker v West Yorkshire Metropolitan Council and Metropolitan Borough of Calderdale (1981) Dig. 1982; 1435**

W slipped and injured herself on an icy patch on the highway. She sued the LHA for negligently allowing water to run onto the highway from land which the authority owned. WYMC relied for notification of defects on the public and on an annual street inspection, but regular inspections were not made and street inspectors were not employed. The court decided that on the facts W could not prove where the water forming the ice had come from. The court said obiter dicta that in the absence of evidence on the availability of resources and the practice of other authorities it was not negligent for the LHA not to have a system of defects inspections or to employ street inspectors.

116   **B L Holdings Ltd v Robert J Wood and Partners (1979) 12 BLR 1**

The plaintiff was a client to the defendants, a group of architects. The plaintiffs asked the defendants about the need for planning permission for offices in Brighton. They were told an Office Development Permit would not be needed if the area of floor space whose sole or principal use was for office purposes did not exceed 10000 ft^2 floor area, 10000 ft^2 of which were for office purposes. When the building was built it could not be used as it was decided an Office Development Permit was needed (ie floor area included the car park). The plaintiffs sued the architects for breach of duty and negligence for not having told them to obtain legal advice concerning the permit. The trial judge found the defendants liable. (A Chief Estates Surveyor employed by the plaintiffs after the offices were built believed it common practice not to include floor area occupied by ancillary services).

The defendants were not negligent or in breach of their duty for believing that the decision about the calculation of floor space for a permit was a matter of opinion or was for the discretion of the local authority.

The architect's duties to the client do not require him to advise the client to seek legal advice on aspects of law which are not clearcut, but for which there is an accepted practice to follow.

Architects need to be aware of differences between common practice and law, and to account for possible loss to the client if common practices are not followed.

117 Hedley Byrne & Co Ltd v Heller and Partners Ltd (1963) 2 ALL ER (HL)

HB who were in contract with Easipower Ltd became doubtful regarding E's financial standing and contacted H & P who were E's bankers, for a reference. H & P replied that E was a '...respectably constituted company considered good for its ordinary business engagement.' HB relied on this statement and loaned money to E. E went into liquidation and HB lost £17,000; HB is attempting to obtain compensation for this loss from H & P. The importance of this case rests on the idea that a banker (and other professionals by analogy), when offering an opinion, owes a duty of care to the party to whom it is being given i.e. to give the opinion in a non-negligent manner after careful consideration, not 'off the cuff'.

In the instant case, whilst the possibility of the duty to exist was established, the duty did not exist by reason of HB's letter of request commencing with a statement that they would accept the opinion without '...legal responsibility on the part of H & P', i.e. it was a legal disclaimer.

118 Clay v Crump and Sons Ltd and others 4 BLR CA

A building was demolished, leaving one wall standing which was to be incorporated into the new building. A discussion between the two contractors and the architect determined that the wall was safe, but a reasonably careful examination would have led to the opposite conclusion. Neither the architect nor the contractors examined the wall carefully and a builders cabin was placed by the wall. The wall collapsed and two men in the hut were killed. The contractor as the employer of the men was liable, he owed a duty to provide a safe system of work, but also the architect and demolition contractor were liable because the two men were '...within the class (group not social class) of person whom they should have had in contemplation as being affected by their acts or omissions and, on the facts, they were in breach of that duty of care...' which was owed to them. The damages awarded were: architect 42%, demolition contractor 38% and the builder 20%.

Appendix 1

Problems and Questions

1. Distinguish between an 'Improvement Line' and a 'Building Line'.

2. Outline the alternative procedures which a developer may follow to ensure the adoption by the local authority of any estate roads which he builds.

3. To enable the efficient clearing away of debris from a small site, the contractor orders a builder's skip from a local hire firm.

   The hire firm park the skip on the highway, ignoring a request of the contractor to park it on the driveway to the site.

   Some time later the contractor is approached by a man in civilian clothes who produces a police warrant card and requests the contractor to move the skip as he considers it to be an obstruction on the highway and a danger to traffic.

   The contractor says 'I would like to cooperate but I didn't park it on the road and anyway, I am not the owner, so I can't move it.' The policeman says he will report the contractor and that a prosecution will most certainly follow.

   Explain the regulations governing the use of a builder's skip. Discuss this particular contractor's problem and possible defences he might invoke if he were prosecuted.

4. A contractor hired a waste disposal hopper and the terms of the agreement were that the contractor would obtain the necessary permit for it to be placed on the highway and would provide the necessary lighting. He forgot these matters and was charged with depositing the hopper unlawfully on the highway. The contractor contended that he was not there when the hopper was delivered and that he neither signed the delivery

documents nor directed the siting of the hopper. Briefly outline the regulations appertaining to parking hoppers on the highway and indicate the likely outcome of the above circumstances.

5. Discuss the awareness the contractor should show when working near the highway and use case law and legislation as a basis for formulating working policy on procedures when involved in operations on or adjacent to the highway.

6. Discuss the legal implications of the following incidents on the same site:

   (i) A builder constructs a hoarding around a site in a town centre and because the building is to be constructed up to the boundary of the site, the hoarding encroaches on to the public footpath.

   (ii) 'Big Daddy' is angry when his little boy whom he is carrying on his shoulders grabs hold of barbed wire put on top of the hoarding to deter trespassers.

   (iii) Mr Dunstan, a blind piano tuner, on his way to work falls over and hurts himself after stumbling into a package of bricks placed on the footpath prior to being taken inside.

   (iv) A lorry leaving site spreads mud all over the highway and when turning runs over the public footpath breaking a kerb stone and cracking several paving slabs.

7. During the course of a refurbishment of a city centre store, the building is shrouded in scaffolding and hoardings which are erected off the footpath. Neighbouring shopkeepers complain that as a result of the work they are losing custom and that they are considering suing both the store and the builder for compensation. They are also going to complain to the council to get something done about it.

   Discuss the legal background to these circumstances and offer some reasoned advice to the builder.

8. When materials are delivered for a domestic extension, on a restricted site, they are deposited on the grass verge by the roadside. Palletted bricks are delivered and placed on the verge but due to the limited space project on to the footpath, reducing its width so that people can pass only in single file.

Whilst the bricks are being unloaded the labour-only bricklayer says, 'We shouldn't block the footpath'. The client (who is pressing for quick completion) says, 'It should be OK. The bricks won't be long there anyway' and the management contractor remarks that, 'As long as people can get through we are not breaking the law'.

When being stacked the bricks slip but do not fall, being held by metal straps.

Later when mothers are bringing the children home from school, Mrs Harassed, pushing a pram with one hand and holding a small child with the other, goes through the restricted footpath and the pram knocks a stack of bricks. Several bricks fall, hitting and injuring the small child.

Discuss the above circumstances with particular reference to the law of Highways and comment on the legal liability of the various parties involved.

9. A developer wishes to develop a private housing scheme on a green field site across which runs a public footpath and bridleway. As the client's professional adviser explain to him the possible restrictions these facts may impose on his freedom of action in planning the layout of the site and outline the legal procedures he may follow to obtain some flexibility in his planning.

10. As a local authority highway engineer explain to the prospective developer of a site the rights, limitation and conditions which may be imposed regarding access to the highway from a prospective development.

Appendix 2

Table of Cases

Case	No.	Page
1	Tarry v Ashton (1876) 1 QBD 314	1,16, 29, 30, 34, 57, 58, 65
2	Ex Parte Lewis (1888) 21 QBD	2, 65
3	Goodtitle and Chester v Alker and Elms (1757) 1 Burr, 146; Vol 21 Hals p	2, 65
4	Hue v Whitely (1929) 1 Ch 440	3, 66
5	Fairey v Southampton CC (1956) 2 QB 439	3, 66
6	A-G v Esher Linoleum Co Ltd (1901) 2 Ch 647	4, 67
7	Williams Ellis v Cobb (1935) 1 KB 310	4, 68
8	Lewis v Thomas (1950) 1 KB 438	5, 68
9	Sandgate UDC v Kent CC (1898) 79 LT HL	5, 68
10	Ramus v Southend Local Board (1892) 67 LT 169	5, 69
11	A G v Blackpool Corporation (1907) 71 JP 478	5, 69
12	Roberts v Webster (1967) 66 LGR 298-305	5, 69
13	Moser v Ambleside UDC (1925) 89 JP 118-120	5, 70
14	Greenwich Board of Works v Maudsley (1870) LR 5QB 397	5, 70
15	Farquhar v Newbury UDC (1909) 1 Ch 12 CA	6, 71
16	A G v Mallock (1931) 146 LT 344	6, 71
17	Cubitt v Lady Caroline Maxse (1873) LR 8 CP 704 Hals Vol 21	6, 72
18	Austins Case (1672) 1 Vent 1879	10, 72
19	Russell v Men of Devon (1788) 2 Term Rep 667	11, 63, 72
20	Thompson v Brighton Corp (1984) 1 QBD 332	11, 73
21	Skilton v Epsom and Ewell UDC (1937) 1 KB 112	11, 73
22	Macclesfield Corp v Great Central Railway (1911) 2 KB 528, CA	12, 74
23	R v Baker (1980) 25 QBD 213	19, 74
24	Rundle v Hearle (1898) 2 QBD 83	25, 74
25	Tunbridge Wells Corporation v Baird (1896) AC 434	14, 15, 75
26	Tottenham UDC v Rowley (1912) 2 Ch 633 CA	14, 19, 75
27	Marriot v East Grinstead Gas and Water Co. (1909) 1 Ch 70	15, 76

28   Wood v Ealing Tenants (1907) 2 KB 390   15, 76

29   Porter v Ipswich Corporation (1922) 2 KB 145,Hals   15, 77

30   Fenna v Clare (1895) 1 QB 199   16, 29, 77

31   Harold v Watney (1898) 2QB 320 CA   16, 78

32   Bromley v Mercer (1922) 2 KB 126, at 131, CA   16, 78

33   Caminer v Northern Investment Trust (1951) AC 88   16, 59, 79

34   Leasne v Lord Egerton (1943) 1 ALL ER 489   16, 30, 79

35   Crane v South Suburban Gas Company (1916)
       1 KB 33   17, 35, 79

36   Halsey v Esso (1961) 2 ALL ER 145   17, 80

37   Lagan Case (1927) AC 226   17, 81

38   Dovaston v Payne (1975) Hals Vol 25   17, 81

39   Hickman v Maisey (1900) 1 QB 752   18, 82

40   Harrison v Duke of Rutland (1893) 1 QB 142   18, 82

41   Rogers v Ministry of Transport (1952) 1 ALL ER 634 18, 27, 83

42   R v Surrey CC Ex p Send Parish Council (1979) 40
       P&CR 390   18, 83

43   Noble v Harrison (1926) 1 KB, 332   16, 59, 84

44   Marshall v Blackpool Corporation (1935) AC16; (1934)
       ALL ER   19, 84

45   Cobb v Saxby (1914) 3 KB 822   19, 84

46   Perry v Stanborough (Developments) Ltd and
       Wimborne DC and Dorset CC (1977) 244 E.G. 551;
       Digest CLY 2502   19, 85

47   Barber v Penley (1893) 2 Ch 447   19, 30, 85

48   Leonidis v Thames Water Authority (1979) 77 LGR 722 19, 85

49   Lyons, Sons & Co v Gulliver (1914) 1 Ch 631 CA   20, 30, 86

50   Harper v Haden (l933) Ch 298   20, 87

51   Millward v Reddith Local Board of Health (1873) 21
       WR 429 Digest (Repl.) 351   20, 87

52   Hawkins v Minister of Housing and Local Government
       (1962) 14 P&CR 44   21, 88

53   Wolverton UDC v Willis (1962) 1 ALL ER 243   28, 88

54   Seekings v Clarke (1961) 59 LGR 269   28, 89

55   Lowdens v Keaveney (1903) 2 IR 82   28, 89

56   Castle v St Augustine's Links (1922) 38 TLR 6l5   29, 89

57   Bolton v Stone (1951) AC 650   29, 90

58   Millar v Jackson (1977) 3 WLR 20; Q.B.966   29, 91

59   Wilson v Kingston upon Thames Corp (1949)
       1 ALL ER 679   11, 91

60   Padbury v Holliday and Greenwood (1912) 28
       TLR   29, 58, 61, 92

61   Reedie v London and North Western Railway Co. (1849) 4,
       Exch 244   29, 58, 92

62   Salsbury v Woodland and Others (1976) 1 QB 324   58, 93

63   Barker v Herbert (1911) 2 KB 633   57, 93

64	Wringe v Cohen (1939) 1 KB 229	58, 94
65	Dymond v Pearce (1972) 1 QB 496	30, 94
66	Haley v London Electricity Board (1965) AC 778, (1964) 3 ALL ER 185	30, 60, 95
67	Ellis v Sheffield Gas Consumers (1853) 2 E&B 767	30, 95
68	Holliday v National Telephone Co (1899) 2QB 392 CA	30, 61, 96
69	Arrowsmith v Jenkins (1963) 1 QB 561	31, 96
70	Lodge Hole Colliery v Wednesbury Comp (1908) AC 323	33, 97
71	Trevett v Lee (1955) 1 ALL ER 406	34, 97
72	Clarke v J Sugrue & Sons (1959) Times, 29 May 1959: CLY 2375	34, 98
73	Stewart v Wright (1893) 9 TLR 480	34, 98
74	AA King (Contractors) v Page (1970) 114,SJ 355	35, 39, 99
75	Gatland v Metropolitan Police (1968) 2 QB 279 DC	35, 99
76	Hunston v Last (1965) 109, SJ 391	35, 100
77	Farrell v Mowlem (1954) 1 Lloyds Rep. 437	34, 100
78	Carshalton UDC v Burrage (1911) 2 ch 133	35, 101
79	Myers v Harrow Corp (1962) 2 QB 442	35, 101
80	Nicholson v Southern Railway Co (1935) 1 KB 558	35, 102
81	New Towns Commission v Hemel Hempstead (1962) 3 ALL ER 183	36, 103
82	Pilling v Abergele UDC (1950) 1 KB 636	36, 103
83	Principality Building Society v Cardiff Corporation (1986) 19 P1CR 821	36, 103
84	Drury v Camden LBC (1972) RTR 391	37, 104
85	Hardcastle v Bielby (1892) IQB 709	37, 104
86	Saper v Hillgate Builders; King v Hillgate Builders (1972) RTR 38 DC	38, 105
87	Lambeth BC v Saunders Transport (1974) RTR 319 DC	38, 106
88	Barnet LBC v S & W Transport (1975) RTR 211	39, 106
89	York City Council v Poller (1976) RTR 37 DC	39, 107
90	Pitman v Southern Electricity Board (1978)3 ALL ER 901	60, 107
91	Wills v T F Martin (Roof Contractors Ltd) (1972) RTR 368	39, 107
92	Derrick v Cornhill (1970) Crim LR 467 CA	39, 108
93	Gabriel v Enfield BC (1971) RTR 265, 69 LGR 382	39, 109
94	Hales Containers Ltd v Ealing LBC (1972) RTR 391	39, 110
95	Westminster Bank v Beverley BC (1961) IQB 488	44, 111
96	Sittingbourne UDC v Liptons Ltd (1931) 1KB 539	44, 111
97	Robinson v The Local Board for the District of Barton-Eccles, Winton and Monton (1883) 8 AC 798	46, 111
98	Devonport Corporation v Tozer (1902) 2 Ch 182	46, 112
99	Astor v Fulham BC (1963) 61 LGR 281; E.G.358	47, 112

100	Buckinghamshire CC v Trigg (1963) 1 ALL ER 403	48, 113
101	Warwickshire CC v Adkins (1976) 66 LGR 486	48, 113
102	Ware v Gaunt (1960) 3 ALL ER 778	48, 50, 114
103	Littler v Liverpool Corp (1968) 2 ALL ER 343	63, 114
104	Sinclair-Lockhart's Trustees v Central Land Board (1951) 1 P&CR	52, 115
105	Whitstable UDC v Campbell (1959) QB JPL 46	53, 56, 116
106	National Employers' Mutual General Insurance Association v Herne Bay UDC (1972) 70 LGR 592	54, 116
107	Griffiths v Liverpool Corp (1966) 2 ALL ER	63, 64, 117
108	Rider v Rider and Another (1973) 2 WLR 190	63, 117
109	Hardaker v Idle District Council (1896) 1 QB 335	64, 117
110	Burnside and Another v Emerson and Others (1968) 3 ALL ER 741	64, 118
111	Haydon v Kent CC (1978) 2 ALL ER 97	64, 118
112	Meggs v Liverpool Corporation (1968) 1 ALL ER 1137	64, 119
113	Burton v West Suffolk CC (1960) 2 ALL ER 226	62, 119
114	PGM Building Co v Kensington and Chelsea Royal Council (1982) RTR 107 DC	39, 119
115	Whitaker v West Yorkshire Metropolitan Council and Metropolitan Borough of Calderdale (1981) Dig. 1982; 1435	63, 120
116	B L Holdings Ltd v Robert J Wood and Partners (1979) 12 BLR 1	61, 120
117	Hedley Byrne v Heller and Partners (1964) AC P475 AC P575 2 ALL ER 1963	62, 121
118	Clay v A J Crump and Sons Ltd and Others (1963) 4 BLR CA (CA 1962)	62, 121

Appendix 3

Table of Statutes

Acquisition of Land (Authorisation Procedure) Act 1946   23

Building Act 1984   8, 32
  S84   9
  S78   32
  S85   9

Building Societies Act 1960   51

Civil Aviation Act 1949
  S28   23

Coast Protection Act 1949
  S 18   8
  S34   8

Countryside Act 1949, 1968   8, 11
  S30   3

Health and Safety at Work etc Act 1974 Part 1   42

Highways Statute of 1563   10

Highways Act 1691   10

Highways Act 1726–1773   6

Highways Act 1835–1885   6, 11, 45

Highways Act 1959   7

Highways (Miscellaneous Provisions) Act 1961   7
  S 1   7
Highways (Miscellaneous Provisions) Act 1971   7

S24   7
Highways Act 1980   7, 8, 19, 23, 26, 27

S10	24	S 118	24
S28	25	S 121	24
S31	3, 5	S 123 (1)	25
S36	12	S 124	22
S38	4, 52, 53, 55	S 127	21
S40	12	S 131	31
S41	61	S 132	33
S51	13	S 133	33
S58	61, 62	S 134	33
S52(2)	62	S 136	7
S73	43	S 137	31
		S 138	35
S73(13)	44	S 139	7, 31, 37
S74	43	S 139(1)	37
S86 8,	60	S 139(3)	38
S 116	23, 24	S 139(4)	38, 40
S 117	24	S 139(5)	39
S 139(9)	37, 38, 39		
S 139(10)	39	8 184	7
S 140(2)	39	S 184(17)	22
S 140(9)	40	S 186	46, 47
S 141	36	S 187	46
S 143	34	S 188	46, 47
S 148	33	S 190(3)	49
S 150	34	S 193	48
S 151	34	S 196	48
S 152	34	S 198	49
S 153	7, 31	S204	49
S 161	35	S205	50
S 162	34	S215	52
S 163	34	S219(4)	51, 54
S 164	34		
S 165	35	S220	53
S 166	35	S221	53
		S222	53
S 168(1)	7	S223(1) (2)	55
S 169	7, 31, 41, 42	S229	53, 55
S 170	36	S230	13
S 171	37	S286	36
S 172	7, 40	S317	24
S 173	31, 40	S322	56
S 175	37	S325	24
S 176	36	S329	46

S 177	36	S329(1)	38, 41, 61
S 178	36	S333	8
S 179	36	S336	8
S 180	36	S338	8
S 184	7, 22	S339	8

Local Government Act 1966	10
Local Government Act 1972	34
Ministry of Transport Act 1919	9
National Parks and Access to the Countryside Act 1949	8, 11

Telegraph Act 1878
S7	8

Town and Country Planning Act 1971 21, 23
S22(1)	8	S214(1) (a)	26
S22	21	S215	26
S23	8	S216	26
S24	49	S221	8
S29	20	S290	8
S76	22	S290(1)	21
S 127	22	S209	8
S210	26	S211	22, 26
S212	8, 26, 27	S212	26
S213	26	S214	26
215	26	216	26

Town Police Clauses Act 1847
S28	31, 34

Transport Act 1919	10
Private Street Works Act 1892	45, 49, 50
Public Health Act 1875	45
S 171(1)	31
Public Utilities (Street Works) Act 1950*	8, 27
Public Works Loans Act 1964	53
Road Traffic Acts	7
Winchester, Statute of 1285	6, 10

Statutory Instruments

Town and Country Planning Regulations 1976 (Reg 14) SI1976
No 1419   27

Construction (Working Places) Regulations 1966 SI1966 No 94   42

General Development Order 1977 SI 1987 No 289   49

Non-Statutory References

J.C.T. '80 cl 6.1.1   32, 41

J.C.T. '80 c120   32

I.C.E. Standard Form of Contract cl 22   59

D.O.E. Circular (9/77 WO circular 8/77)   42

*At the time of publication Parliament is deliberating on the 'New Roads and Street Works Bill' which will amend the Public Utilities (Street Works) Act 1950.

The aims of the new Bill are (a) to allow the private financing of new roads and (b) to regulate excavations by the utilities on the highways more closely than at present.

Bibliography

Halsbury's Laws of England Vol 21

Local Government Law – Davies

Principles of Local Government Law – Cross

General Principles of the Law of torts – James

Street on Torts – Harry Street

An introduction to English Legal History – Baker

Authority on Highway Laws (Winfield & Jolowicz) p. 388, Pratt & Mackenzie, 20th Edition.

Digest of Criminal Law Art.235 Stephen

Index

Access
  right of   19, 21
  infringement of do   19
Adopted   45, 55
Adoption   11, 55, 56
Adjacent owners   14
Advanced Payments Code   51, 53, 54
Appeal   43
Arbitration, statutory   20
Architect   7, 51, 61, 63
Artificial structures   5

Bonfire   35
Boundary   18
Break open highway   7
Bridges   36
Bridleway   2, 3, 10
Builder   51
Building owner   57
Bye-laws   46, 47, 49

Cables   15
Carriage crossing   21
Cellars   36
Churchways   6
Civil liability   11, 41, 57
Classified road   10
Compensation   22, 27, 45
Conditions   21, 37
Constable in uniform   40
Contractor   7, 51, 57, 59, 60
  independent   59
  sub-   51
Construction   47
  programme   33
Control   5

Covered platform	40
Criminal charges	1, 20, 27
Crown Court	44
Cul-de-sacs	5
Damages	17
special	20, 30
Dangerous fencing	35
Dedication	3, 4
statute by	6
Defence	62
Democracy	45, 55
Developer	48, 53
Diversion	7, 23, 24, 25, 26
Drains	36
Driftways	2, 3
Emergency powers	32
Employer (client, building owner)	32
Engineering operations	21
Estimate	50
Exemption	51
Express	4
Extinguish a highway	23, 24, 25
Fences	16, 18
dangerous	35
Financing	46
Firearms	35
Fireworks	35
Foreshore	5
Footpath	2, 3, 10, 20
Frontages	47, 48, 49
Games	35
G.D.O. (General Development Order)	49
Gutters	34, 58
Graffitti	33
Herbage, right of	14
Highways	2
Highways Act 1980	7
definition of	2
creation of	3, 22
law	2
Act 1980	2, 7

authority local 6
authorities 7, 9, 18, 61
administration of Act 7

Highway engineer 50
Hoardings 7, 36, 40, 60
lighting 40, 42
removing 40
licencing 41
Housing estates 45

Implied 4, 23
Indemnity for civil claims 32
Injunction (mandatory) 1, 17, 31

JCT '80 32, 41

Lands Tribunal 22, 43, 48
Latent defects 59
Lines 42
building line 43
compensation 43
frontage line 43
improvement line 43
legal costs 32
Local Acts 6
Local Highway Authority 39
Local Planning Authority 25
London Gazette 22

Magistrates Court 48
Maintenance 7
public, expense 11, 12, 14, 21, 45, 46
private 12
Making up 7, 11
Minister 44, 53
Mixing mortar 7, 36, 37
Money deposited 52
Mud 33

Negligence 59
New streets 45, 46, 47, 48
Nuisance
private 19, 27, 29
abatement of 17
indictment 17

summarily   17

Obligations of adjacent owners   16
Obstruction   20, 27, 28, 31, 41, 59
  wilful   31
  queues   19
Occupiers   57

Passage, rights of   17
Personal injury   57
Planning   49
Planning permission   21, 26
Plans deposited   52
Positioning of buildings   7
Police fund   40
Prescription   3, 12, 23
Private road   11, 54
Private streets   7, 11, 45, 51
Private street works   45, 51, 52, 53, 55
Projections   34
Projections, natural   16
Procedures   49
Public user   6
Public way   54
  unclassified roads   10
  'unmade up' roads   45
  ultra vires   19, 44
  vicarious liability   57
  wires   15

Rights of public   17
Re-siting   21
Resolution   50
Retaining walls   36
Reversing doors and windows   34

Scaffolding   7, 17, 20, 31, 36, 41, 60
Section 38 Agreement   52, 54
Secretary of State for the Environment   25, 26, 49
Serving notice   56
Setting out   6
Shopping mall   26
Shopping precinct   26
Shrubs, planting   35, 36
Skips (Hoppers)   1, 7, 31, 37, 60
  conditions   37, 38, 41

defences	38
fine	37, 49, 51
permission	37
removing	39
repositioning	39
Specification	50
Special roads	7, 9
Spikes on walls	16
Staking materials	20
Statutory offences	31
Statutory undertaker	33, 42
Statute of Westminster	6
Stopping-up	7, 22, 24, 26
Street lighting	50
Strict liability	57
Structures in the street	34
Studs	55
Sub-contractor	51
Surveyor	7, 61, 63
Surveyor of Highways	10, 61, 63
Temporary closing	7
Time Immemorial	3
Trees, decaying	16
Tree roots	14
Trespass	15, 17, 18
Trunk roads	79
Westminster, Statute of	6